MEMOIRS
of the
American Mathematical Society

Number 963

T0074472

Approximate Homotopy of Homomorphisms from $C(X)$ into a Simple C^*-Algebra

Huaxin Lin

May 2010 • Volume 205 • Number 963 (second of 5 numbers) • ISSN 0065-9266

American Mathematical Society
Providence, Rhode Island

Library of Congress Cataloging-in-Publication Data

Lin, Huaxin, 1956-
Approximate homotopy of homomorphisms from $C(X)$ into a simple C^*-algebra / Huaxin Lin.
p. cm. — (Memoirs of the American Mathematical Society, ISSN 0065-9266 ; no. 963)
"Volume 205, number 963 (second of 5 numbers)."
Includes bibliographical references.
ISBN 978-0-8218-5194-4 (alk. paper)
1. Homotopy theory. 2. Homomorphisms (Mathematics). 3. C^*-algebras. I. Title.

QA612.7.L576 2010
514′.24—dc22 2010003517

Memoirs of the American Mathematical Society

This journal is devoted entirely to research in pure and applied mathematics.

Publisher Item Identifier. The Publisher Item Identifier (PII) appears as a footnote on the Abstract page of each article. This alphanumeric string of characters uniquely identifies each article and can be used for future cataloguing, searching, and electronic retrieval.

Subscription information. Beginning with the January 2010 issue, *Memoirs* is accessible from www.ams.org/journals. The 2010 subscription begins with volume 203 and consists of six mailings, each containing one or more numbers. Subscription prices are as follows: for paper delivery, US$709 list, US$567 institutional member; for electronic delivery, US$638 list, US$510 institutional member. Upon request, subscribers to paper delivery of this journal are also entitled to receive electronic delivery. If ordering the paper version, subscribers outside the United States and India must pay a postage surcharge of US$65; subscribers in India must pay a postage surcharge of US$95. Expedited delivery to destinations in North America US$57; elsewhere US$160. Subscription renewals are subject to late fees. See www.ams.org/customers/macs-faq.html#journal for more information. Each number may be ordered separately; *please specify number* when ordering an individual number.

Back number information. For back issues see www.ams.org/bookstore.

Subscriptions and orders should be addressed to the American Mathematical Society, P. O. Box 845904, Boston, MA 02284-5904 USA. *All orders must be accompanied by payment.* Other correspondence should be addressed to 201 Charles Street, Providence, RI 02904-2294 USA.

Copying and reprinting. Individual readers of this publication, and nonprofit libraries acting for them, are permitted to make fair use of the material, such as to copy a chapter for use in teaching or research. Permission is granted to quote brief passages from this publication in reviews, provided the customary acknowledgment of the source is given.

Republication, systematic copying, or multiple reproduction of any material in this publication is permitted only under license from the American Mathematical Society. Requests for such permission should be addressed to the Acquisitions Department, American Mathematical Society, 201 Charles Street, Providence, Rhode Island 02904-2294 USA. Requests can also be made by e-mail to reprint-permission@ams.org.

Memoirs of the American Mathematical Society (ISSN 0065-9266) is published bimonthly (each volume consisting usually of more than one number) by the American Mathematical Society at 201 Charles Street, Providence, RI 02904-2294 USA. Periodicals postage paid at Providence, RI. Postmaster: Send address changes to Memoirs, American Mathematical Society, 201 Charles Street, Providence, RI 02904-2294 USA.

© 2009 by the American Mathematical Society. All rights reserved.
Copyright of individual articles may revert to the public domain 28 years after publication. Contact the AMS for copyright status of individual articles.

This publication is indexed in *Science Citation Index*®, *SciSearch*®, *Research Alert*®, *CompuMath Citation Index*®, *Current Contents*®/*Physical, Chemical & Earth Sciences*.
Printed in the United States of America.

∞ The paper used in this book is acid-free and falls within the guidelines established to ensure permanence and durability.
Visit the AMS home page at http://www.ams.org/

10 9 8 7 6 5 4 3 2 1 15 14 13 12 11 10

Contents

Abstract

In this paper we prove Generalized Homotopy Lemmas. These type of results play an important role in the classification theory of $*$-homomorphisms up to asymptotic unitary equivalence.

Let X be a finite CW complex and let $h_1, h_2 : C(X) \to A$ be two unital homomorphisms, where A is a unital C^*-algebra. We study the problem when h_1 and h_2 are approximately homotopic. We present a K-theoretical necessary and sufficient condition for them to be approximately homotopic under the assumption that A is a unital separable simple C^*-algebra, of tracial rank zero, or A is a unital purely infinite simple C^*-algebra. When they are approximately homotopic, we also give an upper bound for the length of the homotopy.

Suppose that $h : C(X) \to A$ is a monomorphism and $u \in A$ is a unitary (with $[u] = 0$ in $K_1(A)$). We prove that, for any $\epsilon > 0$, and any compact subset $\mathcal{F} \subset C(X)$, there exist $\delta > 0$ and a finite subset $\mathcal{G} \subset C(X)$ satisfying the following: if $\|[h(f), u]\| < \delta$ for all $f \in \mathcal{G}$ and $\text{Bott}(h, u) = 0$, then there exists a continuous rectifiable path $\{u_t : t \in [0,1]\}$ in A such that

$$u_0 = u, \ u_1 = 1_A \text{ and } \|[h(g), u_t]\| < \epsilon \text{ for all } g \in \mathcal{F} \text{ and } t \in [0,1]. \tag{e 0.1}$$

Moreover,

$$\text{Length}(\{u_t\}) \leq 2\pi + \epsilon. \tag{e 0.2}$$

We show that if $\dim X \leq 1$, or A is purely infinite simple, then δ and $\mathcal{G}$ are universal (independent of A or h). In the case that $\dim X = 1$, this provides an improvement of the so-called Basic Homotopy Lemma of Bratteli, Elliott, Evans and Kishimoto for the case that A is as mentioned above. Moreover, we show that δ and $\mathcal{G}$ cannot be universal whenever $\dim X \geq 2$. Nevertheless, we also found that δ can be chosen to be dependent on a measure distribution but independent of A and h. The above version of the so-called Basic Homotopy is also extended to the case that $C(X)$ is replaced by an AH-algebra.

We also present some general versions of the so-called Super Homotopy Lemma.

Received by the editor January 9, 2007.

Article electronically published on December 14, 2009; S 0065-9266(09)00611-5.

2000 *Mathematics Subject Classification.* Primary 46L05, 46L35.

©2009 American Mathematical Society

CHAPTER 1

Prelude

1. Introduction

Let A be a unital C^*-algebra and let $u \in A$ be a unitary which is in the connected component $U_0(A)$ of the unitary group of A containing the identity. Then there is a continuous path of unitaries in $U_0(A)$ starting at u and ending at 1. It is known that the path can be made rectifiable. But, in general, the length of the path has no upper bound. N. C. Phillips ([**46**]) proved that, in any unital purely infinite simple C^*-algebra A, if $u \in U_0(A)$, then the length of the path from u to the identity can be chosen to be smaller than $\pi + \epsilon$ for any $\epsilon > 0$. A more general result proved by the author shows that this holds for any unital C^*-algebras with real rank zero ([**21**]). These basic results have proved to be important in the development in the C^*-algebra theory. This article studies related facts but our results go far deeper. Their impact will be briefly discussed below.

Let X be a path connected finite CW complex. Fix a point $\xi \in X$. Let A be a unital C^*-algebra. There is a trivial homomorphism $h_0 : C(X) \to A$ defined by $h_0(f) = f(\xi)1_A$ for $f \in C(X)$. Suppose that $h : C(X) \to A$ is a unital monomorphism. When can h be homotopic to h_0? When h is homotopic to h_0, how short could the length of the homotopy be? This is one of the questions that motivates this paper.

Let $h_1, h_2 : C(X) \to A$ be two unital homomorphisms. A more general question is when h_1 and h_2 are homotopic? When A is commutative, by the Gelfand transformation, it is a purely topological homotopy question. We only consider noncommutative cases. To the other end of noncommutativity, we only consider the case that A is a unital simple C^*-algebra. To be possible and useful, we actually consider approximate homotopy. So we study the problem when h_1 and h_2 are approximately homotopic. For all applications that we know, the length of the homotopy is extremely important. So we also ask how short the homotopy is if h_1 and h_2 are actually approximately homotopic.

Let X be a path connected metric space. Fix a point ξ_X. Let $Y_X = X \setminus \xi_X$. Let $x \in X$ and let $L(x, \xi_X)$ be the infimum of the length of continuous paths from x to ξ_X. Define

$$L(X, \xi_X) = \sup\{L(x, \xi_X) : x \in X\}.$$

We prove that, for any unital simple C^*-algebra of tracial rank zero, or any unital purely infinite simple C^*-algebra A, if $h : C(X) \to A$ is a unital monomorphism with

$$[h|_{C_0(Y_X)}] = 0 \text{ in } KL(C_0(Y_X), A), \tag{e 1.1}$$

then, for any $\epsilon > 0$ and any compact subset $\mathcal{F} \subset C(X)$, there is a homomorphism $H : C(X) \to C([0,1], A)$ such that

$$\pi_0 \circ H \approx_\epsilon h \text{ on } \mathcal{F}, \quad \pi_1 \circ H = h_0 \text{ and}$$
$$\overline{\text{Length}}(\{\pi_t \circ H\}) \le L(X, \xi_X),$$

where $h_0(f) = f(\xi_X)1_A$ for all $f \in C(X)$ and $\pi_t : C([0,1], A) \to A$ is the point-evaluation at $t \in [0,1]$, and where $\overline{\text{Length}}(\{\pi_t \circ H\})$ is appropriately defined. Note that $[h_0|_{C_0(Y_X)}] = 0$. Thus the condition (e 1.1) is necessary. Moreover, the estimate of length can not be improved.

Suppose that $h_1, h_2 : C(X) \to A$ are two unital homomorphisms. We show that h_1 and h_2 are approximately homotopic if and only if

$$[h_1] = [h_2] \text{ in } KL(C(X), A),$$

under the assumption that A is a unital separable simple C^*-algebra of tracial rank zero, or A is a unital purely infinite simple C^*-algebra. Moreover, we show that the length of the homotopy can be bounded by a universal constant.

Bratelli, Elliott, Evans and Kishimoto ([**3**]) considered the following homotopy question: Let u and v be two unitaries such that u almost commutes with v. Suppose that $v \in U_0(A)$. Is there a rectifiable continuous path $\{v_t : t \in [0,1]\}$ with $v_0 = v$, $v_1 = 1_A$ such that the entire path almost commutes with u? They found that there is an additional obstacle to prevent the existence of such paths of unitaries. The additional obstacle is the Bott element $\text{bott}_1(u, v)$ associated with the pair u and v. They proved what they called the Basic Homotopy Lemma: For any $\epsilon > 0$ there exists $\delta > 0$ satisfying the following: if u, v are two unitaries in a unital separable simple C^*-algebra with real rank zero and stable rank one, or in a unital separable purely infinite simple C^*-algebra A, if $v \in U_0(A)$ and $\text{sp}(u)$ is δ-dense in S^1 except possibly for a single gap,

$$\|uv - vu\| < \delta \text{ and } \text{bott}_1(u, v) = 0,$$

then there exists a rectifiable continuous path of unitaries $\{v_t : t \in [0,1]\}$ such that

$$v_0 = v, \quad v_1 = 1_A \text{ and } \|uv_t - v_tu\| < \epsilon$$

for all $t \in [0,1]$. Moreover,

$$\text{Length}(\{v_t\}) \le 5\pi + 1.$$

Bratteli, Elliott, Evans and Kishimoto were motivated by the study of classification of purely infinite simple C^*-algebras. The Basic Homotopy Lemma played an important role in their work related to the classification of purely infinite simple C^*-algebras and that of [**10**]. It is not surprising that a general Basic Homotopy Lemma may play even more important role in the Elliott program of classification of amenable C^*-algebras. Indeed, the general Basic Homotopy Lemma obtained here (discussed below) leads directly to the results in a subsequent paper ([**35**]) which gives a necessary and sufficient condition in term of KK-theory and traces for which two unital monomorphisms from a unital separable simple amenable C^*-algebra of tracial rank zero with UCT to another such unital simple C^*-algebra are asymptotically unitarily equivalent. This result, together with a recent result of W. Winter ([**55**] and [**36**]), opens the door for more recent advances in the Elliott program (see [**39**], [**52**] and [**38**]). In fact, the general Basic Homotopy Lemma proved here play essential roles in obtaining both uniqueness theorem and existence theorem (the latter may not be entirely expected) required by the recent development (see [**39**]).

It has also been recognized that the Basic Homotopy Lemma is important in the study of AF-embeddings of crossed products ([**45**]). In a subsequent paper ([**37**]) the general Basic Homotopy Lemma plays a key role in solving a version of Voiculescu's problem for AF-embeddings of crossed products: Let X be a compact metric space and let Γ be a $\mathbb{Z}^k$-action on X (for some integer $k \geq 1$). Then the crossed product $C(X) \rtimes_\Gamma \mathbb{Z}^k$ can be embedded into a unital simple AF-algebra if and only if there is a strictly positive Γ-invariant Borel probability measure ([**37**]).

In this paper, we first replace the unitary u in the Basic Homotopy Lemma by a monomorphism $h : C(X) \to A$, where X is a path connected finite CW complex X. A Bott element $\mathrm{bott}_1(h, v)$ can be similarly defined. We prove that, with the assumption that A is a unital simple C^*-algebra of real rank zero and stable rank one, or A is a unital purely infinite simple C^*-algebra, the Basic Homotopy Lemma holds for any compact metric space with dimension no more than one. Moreover, in the case that $K_1(A) = \{0\}$, the constant δ does not depend on the spectrum of h (so that the condition on the spectrum of u in the original Homotopy Lemma can be removed). The proof is shorter than that of the original Homotopy Lemma of Bratteli, Elliott, Evans and Kishimoto. Furthermore, we are able to cut the length of homotopy by more than half (see 3.7 and 11.3).

For a more general compact metric space, the Bott element has to be replaced by a more general map $\mathrm{Bott}(h, v)$. Even with vanishing $\mathrm{Bott}(h, v)$ and with A having tracial rank zero, we show that the respective statement is false whenever $\mathrm{dim} X \geq 2$. However, if we allow that the constant δ not only depends on X and ϵ but also depends on a measure distribution, then a similar homotopy result holds (see 7.4) for unital separable simple C^*-algebras with tracial rank zero. On the other hand, if A is assumed to be purely infinite simple, then there is no such measure distribution. Therefore, for purely infinite simple C^*-algebras, the Basic Homotopy Lemma holds for any compact metric space with shorter lengths. In fact our estimates on the lengths is $2\pi + \epsilon$ (for the case that A is purely infinite simple as well as for the case that A is a unital separable simple C^*-algebra with tracial rank zero).

Several other homotopy results are also discussed. In particular, a version of the Super Homotopy Lemma (of Bratelli, Elliott, Evans and Kishimoto) for finite CW complexes X is also presented.

The presentation is organized as follows:

In section 2, we provide some conventions and a number of facts which will be used later. The definition of $\mathrm{bott}_1(-,-)$ and that of $\mathrm{Bott}(-,-)$ are also given there (2.10).

In section 3, we present the Basic Homotopy Lemma for X being a compact metric space with covering dimension no more than one under the assumption that A is a unital simple C^*-algebra of real rank zero and stable rank one, or A is a unital purely infinite simple C^*-algebra. The improvement is made not only on the upper bound of the length but is also made so that the constant δ does not depend on the spectrum of the homomorphisms (at least for the case that $K_1(A) = 0$).

In section 4 and section 5, we present some results which are preparations for later sections.

In section 6, 7 and 8, we prove a version of the Basic Homotopy Lemma for general compact metric spaces under the assumption that A is a unital separable simple C^*-algebra of tracial rank zero. The lengthy proof is due partly to the complexity caused by our insistence that the constant δ should not be dependent on homomorphisms or A but only on a measure distribution.

In section 9, we show why the constant δ can not be made universal as in the dimension 1 case. A hidden topological obstacle is revealed. We show that the original version of the Basic Homotopy Lemma fails whenever X has dimension at least two for simple C^*-algebras with real rank zero and stable rank one.

In section 10, we present some familiar results about purely infinite simple C^*-algebras.

In section 11, we show that the Basic Homotopy Lemma holds for general compact metric spaces under the assumption that A is a unital purely infinite simple C^*-algebra.

In section 12, we discuss the length of homotopy. A definition related to Lipschitz functions is given there and some elementary facts are also given.

In section 13, we show two homomorphisms are approximately homotopic when they induce the same KL element under the assumption that A is a unital separable simple C^*-algebra of tracial rank zero, or A is a unital purely infinite simple C^*-algebra. We also give an estimate on the upper bound of the length of the homotopy.

In section 14, we extend the results in section 13 to maps which are not necessarily homomorphisms.

In section 15, we present a version of the so-called Super Homotopy Lemma for unital purely infinite simple C^*-algebras.

In section 16, we show that the same version of the Super Homotopy Lemma remains valid for unital separable simple C^*-algebra of tracial rank zero.

In section 17, we show that the Basic Homotopy Lemma in sections 8 and 11 remains valid if we replace $C(X)$ by a unital AH-algebra.

In section 18, we end this paper with a few concluding remarks.

Acknowledgment Most of this research was done in the summer 2006 when the author was in East China Normal University where he had a nice office and necessary computer equipments. It is partially supported by Changjiang Professorship and Shanghai Priority Academic Disciplines. The work is also partially supported by NSF grant (00355273) and (0754813).

2. Conventions and some facts

2.1. Let A be a C^*-algebra. Using notation introduced by [**9**], we denote

$$\underline{K}(A) = \bigoplus_{i=0,1} (K_i(A) \bigoplus \bigoplus_{n>1} K_i(A, \mathbb{Z}/n\mathbb{Z})).$$

Let $m \geq 1$ be an integer. Denote by C_m a commutative C^*-algebra with $K_0(C_m) = \mathbb{Z}/m\mathbb{Z}$ and $K_1(C_m) = 0$. So $K_i(A, \mathbb{Z}/m\mathbb{Z}) = K_i(A \otimes C_m)$, $i = 0, 1$.

A theorem of Dadarlat and Loring ([**9**]) states that

$$Hom_\Lambda(\underline{K}(A), \underline{K}(B)) \cong KL(A, B),$$

if A satisfies the Universal Coefficient Theorem and B is σ-unital (see [**9**] for the definition of $Hom_\Lambda(\underline{K}(A), \underline{K}(B))$). We will identify these two objects.

Let $m \geq 1$ be an integer. Put

$$F_m \underline{K}(A) = \bigoplus_{i=0,1} (K_i(A) \bigoplus \bigoplus_{k|m} K_i(A, \mathbb{Z}/k\mathbb{Z})).$$

2.2. Let B_n be a sequence of C^*-algebras. Denote by $l^\infty(\{B_n\})$ the product of $\{B_n\}$, i.e., the C^*-algebra of all bounded sequences $\{a_n : a_n \in B_n\}$. Denote by $c_0(\{B_n\})$ the direct sum of $\{B_n\}$, i.e, the C^*-algebra of all sequences $\{a_n : a_n \in B_n\}$ for which $\lim_{n\to\infty} \|a_n\| = 0$. Denote by $q_\infty(\{B_n\}) = l^\infty(\{B_n\})/c_0(\{B_n\})$ and by $q : l^\infty(\{B_n\}) \to q_\infty(\{B_n\})$ the quotient map.

2.3. Let A be a C^*-algebra and let B be another C^*-algebra. Let $\epsilon > 0$ and $\mathcal{G} \subset A$ be a finite subset. We say that a contractive completely positive linear map $L : A \to B$ is δ-$\mathcal{G}$-multiplicative if

$$\|L(ab) - L(a)L(b)\| < \delta \text{ for all } a, b \in \mathcal{G}.$$

Denote by $\mathbf{P}(A)$ the set of projections and unitaries in

$$M_\infty(\tilde{A}) \cup \bigcup_{m\geq 1} M_\infty(\widetilde{A \otimes C_m}).$$

We also use L for the map $L \otimes \mathrm{id}_{C_m \otimes M_k} : A \otimes C_m \otimes M_k \to B \otimes C_m \otimes M_k$, $k = 1, 2, ...,$. As in 6.1.1 of [**29**], for a fixed $p \in \mathbf{P}(A)$, if L is δ-$\mathcal{G}$-multiplicative with sufficiently small δ and sufficiently large $\mathcal{G}$, $L(p)$ is close to a projection (with the norm difference less than $1/2$) which will be denoted by $[L(p)]$. Note that if two projections are both close to $L(p)$ within $1/2$, they are equivalent.

If $L : A \to B$ is δ-$\mathcal{G}$-multiplicative, then there is a finite subset $\mathcal{Q} \subset \mathbf{P}(A)$, such that $[L](x)$ is well defined for $x \in \overline{\mathcal{Q}}$, where $\overline{\mathcal{Q}}$ is the image of $\mathcal{Q}$ in $\underline{K}(A)$, which means that if $p_1, p_2 \in \mathcal{Q}$ and $[p_1] = [p_2]$, then $[L(p_1)]$ and $[L(p_2)]$ defines the same element in $\underline{K}(B)$. Moreover, if $p_1, p_2, p_1 \oplus p_2 \in \mathcal{Q}$, $[L(p_1 \oplus p_2)] = [L(p_1)] + [L(p_2)]$ (see 0.6 of [**25**] and 4.5.1 and 6.1.1 of [**29**]). This finite subset $\mathcal{Q}$ will be denoted by $\mathcal{Q}_{\delta,\mathcal{G}}$. Let $\mathcal{P} \subset \underline{K}(A)$. We say $[L]|_\mathcal{P}$ is well defined, if $\overline{\mathcal{Q}_{\delta,\mathcal{G}}} \supset \mathcal{P}$. In what follows, whenever we write $[L]|_\mathcal{P}$, we mean that $[L]|_\mathcal{P}$ is well defined (see also 2.4 of [**7**] for further explanation).

The following proposition is known and has been implicitly used many times.

Proposition 2.4. *Let A be a separable C^*-algebra for which $K_i(A)$ is finitely generated (for $i = 0, 1$), and let $\mathcal{P} \subset \underline{K}(A)$ be a finite subset. Then, there is $\delta > 0$ and a finite subset $\mathcal{G} \subset A$ satisfying the following: If B is a unital C^*-algebra and if $L : A \to B$ is a δ-$\mathcal{G}$-multiplicative contractive completely positive linear map, there is an element $\kappa \in Hom_\Lambda(\underline{K}(A), \underline{K}(B))$ such that*

$$[L]|_\mathcal{P} = \kappa|_\mathcal{P}. \tag{e 2.1}$$

Moreover, there is a finite subset $\mathcal{P}_A \subset \underline{K}(A)$ such that, if $[L]|_{\mathcal{P}_A}$ is well defined, there is a unique $\kappa \in Hom_\Lambda(\underline{K}(A), \underline{K}(B))$ such that (e 2.1) *holds.*

Proof. Since $K_i(A)$ is finitely generated ($i = 0, 1$), by 2.11 of [**9**],

$$Hom_\Lambda(\underline{K}(A), \underline{K}(B)) = Hom_\Lambda(F_m\underline{K}(A), F_m\underline{K}(B))$$

for some $m \geq 1$. Thus it is clear that it suffices to show the first part of the proposition. Suppose that the first part of the lemma fails. One obtains a finite subset $\mathcal{P} \subset \underline{K}(A)$, a sequence of σ-unital C^*-algebras B_n, a sequence of positive numbers $\{d_n\}$ with $\sum_{n=1}^{\infty}\delta_n<\infty$, a finite subsets $\mathcal{G}_n\subset A$ with $\cup_{n=1}^{\infty}\mathcal{G}_n$ is dense in A, and a sequence of δ_n-$\mathcal{G}_n$-multiplicative contractive completely positive linear maps $L_n: A\to B_n$ such that there exists no $\kappa\in Hom_\Lambda(\underline{K}(A),\underline{K}(B))$ satisfying (e 2.1).

Define $\Phi: A\to l^\infty(\{B_n\otimes\mathcal{K}\})$ by $\Phi(a)=\{L_n(a)\}$ for $a\in A$ and define $\bar{\Phi}: A\to q_\infty(\{B_n\otimes\mathcal{K}\})$ by $\bar{\Phi}=\pi\circ\Phi$, where $\pi: l^\infty(\{B_n\otimes\mathcal{K}\})\to q_\infty(\{B_n\otimes\mathcal{K}\})$ is the quotient map. Thus we obtain an element $\alpha\in Hom_\Lambda(F_m\underline{K}(A),F_m\underline{K}(q_\infty(\{B_n\otimes\mathcal{K}\})))$ such that $[\bar{\Phi}]=\alpha$. Since $K_i(A)$ is finitely generated $(i=0,1)$, by 2.11 of [**9**], there is an integer $m\ge 1$ such that

$$Hom_\Lambda(\underline{K}(A),\underline{K}(q_\infty(\{B_n\otimes\mathcal{K}\})))\cong Hom_\Lambda(F_m\underline{K}(A),F_m\underline{K}(q_\infty(\{B_n\otimes\mathcal{K}\}))\ \text{and}$$
$$Hom_\Lambda((\underline{K}(A),\underline{K}(B_n))\cong Hom_\Lambda(F_m\underline{K}(A),F_m\underline{K}(B_n)).$$

By applying 7.2 of [**32**] and the proof of 7.5 of [**32**], for all larger n, there is an element

$$\kappa_n\in Hom_\Lambda(F_m\underline{K}(A),F_m\underline{K}(B_n))$$

such that

$$[L_n]|_{\mathcal{P}}=\kappa_n|_{\mathcal{P}}.$$

This contradicts the assumption that the first part of the lemma fails. □

The following is well known and follows immediately from the definition.

Proposition 2.5. *Let A be a unital amenable C^*-algebra. For any finite subset $\mathcal{P}\subset\underline{K}(A)$, there exists $\delta>0$ and a finite subset $\mathcal{G}\subset A$ satisfying the following: for any pair of δ-$\mathcal{G}$-multiplicative contractive completely positive linear maps $L_1,L_2: A\to B$ (for any unital C^*-algebra B),*

$$[L_1]|_{\mathcal{P}}=[L_2]|_{\mathcal{P}}$$

provided that

$$L_1\approx_\delta L_2\ \text{on}\ \mathcal{G}.$$

2.6. Let B be a C^*-algebra and $C=C([0,1],B)$. Define $\pi_t: C\to B$ by $\pi_t(f)=f(t)$ for all $f\in C$. This notation will be used throughout this article.

The following follows immediately from 2.5 and will be used frequently without further notice.

Proposition 2.7. *Let A and B be two unital C^*-algebras and let $L: A\to B$ be a contractive completely positive linear map. Let $\mathcal{Q}\subset\mathbf{P}(A)$ be a finite subset. Suppose that, for some small $\delta>0$ and a large finite subset $\mathcal{G}$, L is δ-$\mathcal{G}$-multiplicative and $\mathcal{Q}_{\delta,\mathcal{G}}\supset\mathcal{Q}$. Put $\mathcal{P}=\overline{\mathcal{Q}}$ in $\underline{K}(A)$. Suppose that $H: A\to C([0,1],B)$ is a contractive completely positive linear map such that $\pi_0\circ H=L$ and $\pi_t\circ H$ is δ-$\mathcal{G}$-multiplicative for each $t\in[0,1]$. Then, for each $t\in[0,1]$,*

$$[\pi_t\circ H]|_{\mathcal{P}}=[L]|_{\mathcal{P}}.$$

The following follows immediately from 2.1 of [**40**].

Lemma 2.8. *Let B be a separable amenable C^*-algebra. For any $\epsilon>0$ and any finite subset $\mathcal{F}_0\subset B$ there exists a finite subset $\mathcal{F}_1\subset B$ and $\delta>0$ satisfying the following: Suppose that A is a unital C^*-algebra, $\phi: B\to A$ is a unital homomorphism and $u\in A$ is a unitary such that*

$$\|[\phi(a),u]\|<\delta\ \text{for all}\ a\in\mathcal{F}_1. \tag{e 2.2}$$

Then there is an ϵ-$\mathcal{F}_0 \otimes S$-multiplicative contractive completely positive linear map $\psi : B \otimes C(S^1) \to A$ such that

$$\|\psi(a \otimes g) - \phi(a)g(u)\| < \epsilon \tag{e 2.3}$$

for all $a \in \mathcal{F}_0$ and $g \in S$, where $S = \{1_{C(S^1)}, z\}$ and $z \in C(S^1)$ is the standard unitary generator of $C(S^1)$.

2.9. Let A be a unital C^*-algebra. Denote by $U(A)$ the group of all unitaries in A. Denote by $U_0(A)$ the path connected component of $U(A)$ containing 1_A.

Denote by $\mathrm{Aut}(A)$ the group of automorphisms on A. If $u \in U(A)$, denote by $\mathrm{ad}\, u$ the inner automorphism defined by $\mathrm{ad}\, u(a) = u^*au$ for all $a \in A$.

Definition 2.10. Let A and B be two unital C^*-algebras. Let $h : A \to B$ be a unital homomorphism and $v \in U(B)$ such that

$$h(g)v = vh(g) \text{ for all } g \in A.$$

Thus we obtain a homomorphism $\bar{h} : A \otimes C(S^1) \to B$ by $\bar{h}(f \otimes g) = h(f)g(v)$ for $f \in A$ and $g \in C(S^1)$. Put $SA = C_0((0,1), A)$ and identify $A \otimes C(S^1)$ with $C(S^1, A)$. From the following splitting exact sequence:

$$0 \to SA \to A \otimes C(S^1) \leftrightarrows A \to 0 \tag{e 2.4}$$

and the isomorphisms $K_i(A) \to K_{1-i}(SA)$ $(i = 0, 1)$ given by the Bott periodicity, one obtains two injective homomorphisms:

$$\beta^{(0)} : K_0(A) \to K_1(A \otimes C(S^1)) \tag{e 2.5}$$

$$\beta^{(1)} : K_1(A) \to K_0(A \otimes C(S^1)). \tag{e 2.6}$$

Note, in this way, by (e 2.4), that one can write $K_i(A \otimes C(S^1)) = K_i(A) \oplus \beta^{(1-i)}(K_{1-i}(A))$. We use $\widehat{\beta^{(i)}} : K_i(A \otimes C(S^1)) \to \beta^{(1-i)}(K_{1-i}(A))$ for the projection to $\beta^{(1-i)}(K_{1-i}(A))$.

For each integer $k \geq 2$, one also obtains the following injective homomorphisms:

$$\beta_k^{(i)} : K_i(A, \mathbb{Z}/k\mathbb{Z})) \to K_{1-i}(A \otimes C(S^1), \mathbb{Z}/k\mathbb{Z}), i = 0, 1. \tag{e 2.7}$$

Thus we write

$$K_{1-i}(A \otimes C(S^1), \mathbb{Z}/k\mathbb{Z}) = K_{1-i}(A, \mathbb{Z}/k\mathbb{Z}) \bigoplus \beta_k^{(i)}(K_i(A, \mathbb{Z}/k\mathbb{Z})),\ i = 0, 1. \tag{e 2.8}$$

Denote by $\widehat{\beta_k^{(i)}} : K_i(A \otimes C(S^1), \mathbb{Z}/k\mathbb{Z}) \to \beta_k^{(1-i)}(K_{1-i}(A, \mathbb{Z}/k\mathbb{Z}))$ similarly to that of $\widehat{\beta^{(i)}}$., $i = 1, 2$. If $x \in \underline{K}(A)$, we use $\boldsymbol{\beta}(x)$ for $\beta^{(i)}(x)$ if $x \in K_i(A)$ and for $\beta_k^{(i)}(x)$ if $x \in K_i(A, \mathbb{Z}/k\mathbb{Z})$. Thus we have a map $\boldsymbol{\beta} : \underline{K}(A) \to \underline{K}(A \otimes C(S^1))$ as well as $\widehat{\boldsymbol{\beta}} : \underline{K}(A \otimes C(S^1)) \to \boldsymbol{\beta}(\underline{K}(A))$. Thus one may write $\underline{K}(A \otimes C(S^1)) = \underline{K}(A) \bigoplus \boldsymbol{\beta}(\underline{K}(A))$.

On the other hand $\bar{h}$ induces homomorphisms $\bar{h}_{*i,k} : K_i(A \otimes C(S^1)), \mathbb{Z}/k\mathbb{Z}) \to K_i(B, \mathbb{Z}/k\mathbb{Z})$, $k = 0, 2, ...,$ and $i = 0, 1$.

We use $\mathrm{Bott}(h, v)$ for all homomorphisms $\bar{h}_{*i,k} \circ \beta_k^{(i)}$. We write

$$\mathrm{Bott}(h, v) = 0,$$

if $\bar{h}_{*i,k} \circ \beta_k^{(i)} = 0$ for all $k \geq 1$ and $i = 0, 1$.

We will use $\mathrm{bott}_1(h, v)$ for the homomorphism $\bar{h}_{1,0} \circ \beta^{(1)} : K_1(A) \to K_0(B)$, and $\mathrm{bott}_0(h, u)$ for the homomorphism $\bar{h}_{0,0} \circ \beta^{(0)} : K_0(A) \to K_1(B)$.

Since A is unital, if $\mathrm{bott}_0(h, v) = 0$, then $[v] = 0$ in $K_1(B)$.

In what follows, we will use z for the standard generator of $C(S^1)$ and we will often identify S^1 with the unit circle without further explanation. With this identification z is the identity map from the circle to the circle.

Now let $A = C(S^1)$ and $u = h(z)$. Put $\mathrm{bott}_1(u,v) = \mathrm{bott}_1(h,u)([z])$. Note that, if $[v] = 0$ in $K_1(B)$, then $\mathrm{bott}_0(h,v) = 0$. In this case, $K_i(C(S^1))$ is free, thus $\bar{h}_{*i,k} = 0$, if $k \geq 2$. In particular, if $[v] = 0$ in $K_1(B)$ and $\mathrm{bott}_1(h,v) = 0$, then

$$\mathrm{Bott}(h,v) = 0.$$

Suppose that $\{v_n\}$ is a sequence of unitaries in A such that

$$\lim_{n\to\infty} \|h(a)v_n - v_n h(a)\| = 0 \text{ for all } a \in A.$$

Then we obtain a sequential asymptotic morphism $\psi_n : A \otimes S^1 \to B$ such that

$$\lim_{n\to\infty} \|\psi_n(a \otimes g) - h(a)g(v_n)\| = 0 \quad \text{for all } a \in A \text{ and } g \in C(S^1)$$

(see 2.8).

Fixing a finite subset $\mathcal{P} \subset \underline{K}(A)$, for sufficiently large n, as in 2.4, $[\psi_n]|_{\beta(\mathcal{P})}$ is well defined. We will denote this by

$$\mathrm{Bott}(h, v_n)|_{\mathcal{P}}.$$

In other words, for a fixed finite subset $\mathcal{P} \subset \underline{K}(A)$, there exists $\delta > 0$ and a finite subset $\mathcal{G} \subset A$ such that, if $v \in B$ is a unitary for which

$$\|h(a)v - vh(a)\| < \delta \text{ for all } a \in \mathcal{G},$$

then $\mathrm{Bott}(h,v)|_{\mathcal{P}}$ is well defined. In what follows, whenever we write $\mathrm{Bott}(h,v)|_{\mathcal{P}}$, we mean that δ is sufficiently small and $\mathcal{G}$ is sufficiently large so it is well defined. If $\mathcal{P} \subset K_i(A)$, we will write

$$\mathrm{bott}_i(h,v)|_{\mathcal{P}} = \mathrm{Bott}(h,v)|_{\mathcal{P}}, \ i = 0, 1. \tag{e 2.9}$$

Now suppose that A is also amenable and $K_i(A)$ is finitely generated $(i = 0,1)$. For example, $A = C(X)$, where X is a finite CW complex. So, for all sufficiently large n, ψ_n defines an element $[\psi_n]$ in $KL(A,B)$ (see 2.4). Therefore, for a fixed finite subset $\mathcal{P}_0 \subset \underline{K}(A)$, there exists $\delta_0 > 0$ and a finite subset $\mathcal{G}_0 \subset A$ such that, if $v \in B$ is a unitary for which

$$\|h(a)v - vh(a)\| < \delta_0 \text{ for all } a \in \mathcal{G}_0,$$

then $\mathrm{Bott}(h,v)|_{\mathcal{P}_0}$ is well defined. Thus when $K_i(A)$ is finitely generated, $\mathrm{Bott}(h,v)|_{\mathcal{P}_0}$ defines $\mathrm{Bott}(h,v)$ for some sufficiently large finite subset $\mathcal{P}_0$. In particular, $\mathrm{bott}_i(h,v)$ is defined $(i = 0,1)$. In what follows such $\mathcal{P}_0$ may be denoted by $\mathcal{P}_A$. Suppose that $\mathcal{P} \subset \underline{K}(A)$ is a larger finite subset, and $\mathcal{G} \supset \mathcal{G}_0$ and $0 < \delta < \delta_0$.

A fact that we be used in this paper is that, $\mathrm{Bott}(h,v)|_{\mathcal{P}}$ *defines the same map* $\mathrm{Bott}(h,v)$ *as* $\mathrm{Bott}(h,v)|_{\mathcal{P}_0}$ *defines, if*

$$\|h(a)v - vh(a)\| < \delta \text{ for all } a \in \mathcal{G}.$$

In what follows, in the case that $K_i(A)$ is finitely generated, whenever we write $\mathrm{Bott}(h,v)$, we always assume that δ is smaller than δ_0 and $\mathcal{G}$ is larger than $\mathcal{G}_0$ so that $\mathrm{Bott}(h,v)$ is well-defined.

2.11. In the case that $A=C(S^1)$, there is a concrete way to visualize $\mathrm{bott}_1(h,v)$. It is perhaps helpful to describe it here. The map $\mathrm{bott}_1(h,v)$ is determined by $\mathrm{bott}_1(h,v)([z])$, where z is, again, the identity map on the unit circle.

Denote $u=h(z)$ and define

$$f(e^{2\pi it})=\begin{cases}1-2t, & \text{if } 0\le t\le 1/2,\\ -1+2t, & \text{if } 1/2<t\le 1,\end{cases}$$

$$g(e^{2\pi it})=\begin{cases}(f(e^{2\pi it})-f(e^{2\pi it})^2)^{1/2} & \text{if } 0\le t\le 1/2,\\ 0, & \text{if } 1/2<t\le 1 \quad\text{and}\end{cases}$$

$$h(e^{2\pi it})=\begin{cases}0, & \text{if } 0\le t\le 1/2,\\ (f(e^{2\pi it})-f(e^{2\pi it})^2)^{1/2}, & \text{if } 1/2<t\le 1.\end{cases}$$

These are non-negative continuous functions defined on the unit circle. Suppose that $uv=vu$. Define

$$e(u,v)=\begin{pmatrix} f(v) & g(v)+h(v)u^* \\ g(v)+uh(v) & 1-f(v)\end{pmatrix} \tag{e 2.10}$$

Then $e(u,v)$ is a projection. There is $\delta_0>0$ (independent of the unitaries u,v and A) such that if $\|[u,v]\|<\delta_0$, the spectrum of the positive element $e(u,v)$ has a gap at $1/2$. The bott element of u and v is an element in $K_0(A)$ as defined in [**11**] and [**12**] namely

$$\mathrm{bott}_1(u,v)=[\chi_{1/2,\infty}(e(u,v))]-[\begin{pmatrix}1&0\\0&0\end{pmatrix}]. \tag{e 2.11}$$

Note that $\chi_{1/2,\infty}$ is a continuous function on $\mathrm{sp}(e(u,v))$. Suppose that $\mathrm{sp}(e(u,v))\subset(-\infty,a]\cup[1-a,\infty)$ for some $0<a<1/2$. Then $\chi_{1/2,\infty}$ can be replaced by any other positive continuous function F such that $F(t)=0$ if $t\le a$ and $F(t)=1$ if $t\ge 1/2$.

2.12. Let X be a finite CW complex and let A be a unital separable stably finite C^*-algebra. Recall that an element $\alpha\in Hom_\Lambda(\underline{K}(C(X)),\underline{K}(A))$ is said to be positive, if

$$\alpha(K_0(C(X))_+\setminus\{0\})\subset K_0(A)_+\setminus\{0\}.$$

These positive elements may be denoted by $KL(C(X),A)_+$.

2.13. Let X be a connected finite CW complex and let $\xi_X\in X$ be a base point. Put $Y_X=X\setminus\{\xi_X\}$.

2.14. Let X be a compact metric space with metric $\mathrm{dist}(-,-)$. In what follows, we will use

$$\mathrm{dist}((x,t),(y,s))=\sqrt{\mathrm{dist}(x,y)^2+|t-s|^2}$$

for $(x,t),(y,s)\in X\times S^1$.

2.15. Let A and B be two C^*-algebras and $\phi:A\to B$ be a contractive positive linear map. Let $\mathcal{G}\subset A$ be a subset and let $\sigma>0$. We say that ϕ is σ-$\mathcal{G}$-injective if

$$\|\phi(a)\|\ge\sigma\|a\|$$

for all $a\in\mathcal{G}$.

2.16. Let X be a compact metric space and let A be a unital C^*-algebra. Suppose that $\phi : C(X) \to A$ is a homomorphism. There is a compact subset $F \subset X$ such that

$$\ker\phi = \{f \in C(X) : f|_F = 0\}.$$

Thus there is a monomorphism $h_1 : C(F) \to A$ such that $h = h_1 \circ \pi$, where $\pi : C(X) \to C(F)$ is the quotient map. We say the spectrum of ϕ is F.

2.17. Let $C = A \otimes B$, where A and B are unital C^*-algebras. Let D be another unital C^*-algebra. Suppose that $\phi : C \to D$ is a homomorphism. Throughout this work, by $\phi|_A : A \to D$ we mean the homomorphism defined by $\phi|_A(a) = \phi(a \otimes 1)$ for all $a \in A$.

2.18. Let $L_1, L_2 : A \to B$ be two maps, let $\epsilon > 0$ and $\mathcal{F} \subset A$ be a subset. We write

$$L_1 \approx_\epsilon L_2 \text{ on } \mathcal{F},$$

if

$$\|L_1(f) - L_2(f)\| < \epsilon \text{ for all } f \in \mathcal{F}.$$

2.19. Let X be a compact metric space, $\delta > 0$ and $\mathcal{F} \subset C(X)$ be a finite subset. Suppose that $d(X)$ is the diameter of X. Let $\sigma = \sigma_{X,\mathcal{F},\delta}$ be the supremum of those $d(X) \geq \sigma' > 0$ which satisfy the following:

$$|f(x) - f(x')| < \delta \text{ for all } f \in \mathcal{F}, \tag{e 2.12}$$

provided $\mathrm{dist}(x, x') < \sigma'$. Note that if $0 < \sigma_1 < \sigma_{X,\mathcal{F},\delta}$, then $\sigma_1 < \sigma'$ for some $\sigma' > 0$ for which (e 2.12) holds. Thus (e 2.12) holds whenever $\mathrm{dist}(x, x') < \sigma_{X,\delta,\mathcal{F}}$. For any compact metric space X, we may always assume that $d(X) < \infty$. In this way, $\sigma_{X,\mathcal{F},\delta} < \infty$.

2.20. Let τ be a state on $C(X)$. Denote by μ_τ the probability Borel measure induced by τ.

2.21. Let X be a metric space and let $x \in X$. Suppose that $r > 0$. We will use $O(x, r)$ for

$$\{y \in X : \mathrm{dist}(x, y) < r\}.$$

2.22. (see 1.2 of [**40**]) Let X be a compact metric space, let A be a unital C^*-algebra and let $L : C(X) \to A$ be a contractive linear map. Suppose $\epsilon > 0$ and $\mathcal{F}$ is a finite subset of $C(X)$. Denote by $\Sigma_\epsilon(L, \mathcal{F})$ the closure of the subset of those points $\lambda \in X$ for which there is a non-zero hereditary C^*-subalgebra B of A satisfying

$$\|(f(\lambda) - L(f))b\| < \epsilon \text{ and } \|b(f(\lambda) - L(f))\| < \epsilon$$

for all $f \in \mathcal{F}$ and $b \in B$ with $\|b\| \leq 1$. Note that if $\epsilon < \sigma$, then $\Sigma_\epsilon(L, \mathcal{F}) \subset \Sigma_\sigma(L, \mathcal{F})$.

Lemma 2.23. *Let X be a compact metric space, let $\epsilon > 0$, $\sigma > 0$ and let $\mathcal{F} \subset C(X)$ be a finite subset of the unit ball. There is $\delta > 0$ and a finite subset $\mathcal{G} \subset C(X)$ satisfying the following: For any unital C^*-algebra A, any unital δ-$\mathcal{G}$-multiplicative contractive completely positive linear map $L : C(X) \to A$, if L is also $1/2$-$\mathcal{G}$-injective, then $\Sigma_\epsilon(L, \mathcal{F})$ is σ-dense in X.*

PROOF. Choose a σ-dense finite subset $\{x_1, x_2, ..., x_m\}$ in X. Let $\eta = \sigma_{X,\mathcal{F},\epsilon/32}$. In particular,

$$|f(x) - f(x')| < \epsilon/3 \text{ for all } f \in \mathcal{F} \tag{e 2.13}$$

whenever $\mathrm{dist}(x, x') < \eta$. Choose non-negative functions $g_1, g_2, ..., g_m, g_1', g_2', ..., g_m'$ $\in C(X)$ such that $g_i(x) \le 1$, $g_i(x) = 1$ if $x \in O(x_i, \eta/4)$ and $g_i(x) = 0$ if $x \notin O(x_i, \eta/2)$ and $g_i'(x) \le 1$, $g_i'(x) = 1$ if $x \in O(x_i, \eta/8)$ and $g_i'(x) = 0$ if $x \in O(x_i, \eta/4)$, $i = 1, 2, ..., m$. Choose $\delta < \epsilon/4$ and

$$\mathcal{G} = \mathcal{F} \cup \{g_1, g_2, ..., g_m, g_1', g_2', ..., g_m'\}.$$

Now suppose that $L : C(X) \to A$ is δ-$\mathcal{G}$-multiplicative and 1/2-$\mathcal{G}$-injective. Let $B_i = \overline{L(g_i')AL(g_i')}$, $i = 1, 2, ..., m$. Since L is 1/2-$\mathcal{G}$-injective, $B_i \ne \{0\}$, $i = 1, 2, ..., m$. Let $b = L(g_i')cL(g_i')$ for some $c \in A$ such that $\|b\| \le 1$. Then, by (e 2.13),

$$\begin{aligned} \|(f(x_i) - L(f))b\| &\le \|(f(x_1) - L(f))L(g_i)L(g_i')cL(g_i')\| + \delta \\ &\le \|L((f(x_i) - f)g_i)b\| + \delta + \delta \\ &< \epsilon/3 + \delta + \delta < 5\epsilon/6. \end{aligned} \tag{e 2.14}$$

It follows that

$$\|(f(x_i) - L(f))b\| < \epsilon$$

for all $b \in B_i$, $i = 1, 2, ..., m$. Similarly,

$$\|b(f(x_i) - L(f))\| < \epsilon$$

for $i = 1, 2, ..., m$. Therefore $x_i \in \Sigma_\epsilon(L, \mathcal{F})$, $i = 1, 2, ..., m$. It follows that $\Sigma_\epsilon(L, \mathcal{F})$ is σ-dense in X. □

Lemma 2.24. *Let X be a compact metric space and let $d > 0$. There is a finite subset $\mathcal{G} \subset C(X)$ satisfying the following:*

For any unital C^*-algebra A and a unital homomorphism $h : C(X) \to A$ which is 1/2-$\mathcal{G}$-injective, the spectrum F of h is d-dense in X.

PROOF. Let $\{x_1, x_2, ..., x_n\}$ be a $d/2$-dense subset of X. Let $f_i \in C(X)$ such that $0 \le f_i \le 1$, $f_i(x) = 1$ if $x \in O(x_i, d/4)$ and $f_i(x) = 0$ if $x \notin O(x_i, d/2)$. Put

$$\mathcal{F} = \{f_1, f_2, ..., f_n\}.$$

Let $\eta = \sigma_{X,\mathcal{F},1/4}$. Let $g_i, g_i' \in C(X)$ be as in the proof of 2.23, $i = 2, ..., n$. Suppose that $h : C(X) \to A$ is a unital homomorphism which is 1/2-$\mathcal{G}$-injective. Let F be the spectrum of h.

If F were not d-dense, then there is i such that $O(x_i, d/2) \cap F = \emptyset$. Then $f_i(x) = 0$ for all $x \in F$. It follows that $h(f_i) = 0$. However, as in the proof of 2.23, we have

$$\|b\| = \|f(x_i)b\| = \|(f(x_i) - h(f))b\| < 1/2$$

for any $b \in B_1$ with $\|b\| = 1$. A contradiction. □

Lemma 2.25. (1) *Let X be a connected finite CW complex and let A be a unital anti-liminal C^*-algebra. Then there exists a unital monomorphism $\phi_0 : C(X) \to A$ and a unital monomorphism $\phi_1 : C([0,1]) \to A$ and a unital monomorphism $\psi : C(X) \to C([0,1])$ such that*

$$\phi_0 = \phi_1 \circ \psi. \tag{e 2.15}$$

(2) *If X is a finite CW complex with k connected components and A is a unital C^*-algebra with k mutually orthogonal projections $p_1, p_2, ..., p_k$ such that each p_iAp_i is anti-liminal, then there is a unital monomorphism $\psi_0 : C(X) \to A$ such that*

$$[\phi_0] = [\psi_{00}] \text{ in } KK(C(X), A),$$

where ψ_{00} is a point-evaluation on k points with one point on each component.

(3) *If X is a compact metric space and if A is a unital simple C^*-algebra with real rank zero, then there is a unital monomorphism $\phi_0 : C(X) \to A$ and a unital monomorphism $\phi_1 : C(\Omega) \to A$ and a unital monomorphism $\psi : C(X) \to C(\Omega)$ such that*

$$\phi_0 = \phi_1 \circ \psi, \tag{e 2.16}$$

where Ω is the Cantor set.

PROOF. It is clear that part (2) of the lemma follows from part (1). So we may assume that X is connected. It is well known that there is a surjective continuous map $s : [0, 1] \to X$. This, in turn, gives a unital monomorphism $\psi : C(X) \to C([0, 1])$. It follows from [**1**] that there is a positive element $a \in A$ such that $sp(a) = [0, 1]$. Define $\phi_1 : C([0, 1]) \to A$ by $\phi_1(f) = f(a)$ for all $f \in C([0, 1])$. Define $\phi_0 = \phi_1 \circ \psi$.

To see part (3), we note that there is unital monomorphism $\psi_0 : C(\Omega) \to A$ and a unital monomorphism $\psi : C(X) \to C(\Omega)$. Then define $\phi_1 = \psi_0 \circ \psi$. □

2.26. Two projections p and q in a C^*-algebra A are siad to be equivalent if there exists $w \in A$ such that $w^*w = p$ and $ww^* = q$.

2.27. Let A be a stably finite C^*-algebra. We will use $T(A)$ for the tracial state space of A. Denote by $Aff(T(A))$ the real continuous affine functions on $T(A)$. Let $\tau \in T(A)$, we will also use τ for $\tau \otimes Tr$ on $A \otimes M_n$, where Tr is the standard trace on M_n.

There is a (positive) homomorphism $\rho_A : K_0(A) \to Aff(T(A))$ defined by $\rho([p]) = \tau(p)$ for any projection $p \in M_n(A)$. Denote by kerρ_A (or kerρ) the kernel of the map ρ_A.

2.28. A unital C^*-algebra A is purely infinite and simple, if $A \neq \mathbb{C}$, and for any $a \neq 0$, there are $x, y \in A$ such that $xay = 1$. We refer the reader to [**5**], [**48**], [**49**], [**46**] and [**47**], for example, for some related results about purely infinite simple C^*-algebras used in this paper.

A unital separable simple C^*-algebra A with tracial rank zero has real rank zero, stable rank one and weakly unperforated $K_0(A)$. When A has tracial rank zero, we write $TR(A) = 0$. We refer the reader to [**27**], [**31**], [**28**], [**30**], [**4**], [**53**] and [**54**] for more information.

Most of this section will not be used until Chapter 2.

3. The Basic Homotopy Lemma for $\dim(X) \leq 1$

In what follows, we use **B** *for the class of unital C^*-algebras which are simple C^*-algebras with real rank zero and stable rank one, or purely infinite simple C^*-algebras.*

3.1. When X is a connected finite CW complex of covering dimension 1, $X \times S^1$ is of dimension 2. Then the following are true.

1) $K_0(C(X)) = \mathbb{Z}$, $K_0(C(X \times S^1)) = K_0(C(X)) \bigoplus \beta_1(K_1(C(X))) = \mathbb{Z} \bigoplus \beta_1(K_1(C(X)))$ and $K_1(C(X \times S^1)) = K_1(C(X)) \bigoplus \beta_0(K_0(C(X))) = K_1(C(X)) \bigoplus \mathbb{Z}$. In particular, $K_i(C(X \times S^1))$ is torsion free and $\mathrm{ker}\rho_{C(X \times S^1)} = \beta_1(K_1(C(X)))$. Since $K_i(C(X \times S^1))$ is torsion free and finitely generated, one has that $KL(C(X \times S^1), A) = KK(C(X \times S^1), A) = Hom(K_*(C(X \times S^1)), K_*(A))$ for any unital C^*-algebra A.

2) Let A be a unital C^*-algebra. Suppose that $h : C(X) \to A$ is a unital homomorphism and $u \in A$ is a unitary. Fix a finite subset $\mathcal{F}_0 \subset C(X)$ and $\epsilon > 0$. Suppose that $\|uh(a) - h(a)u\| < \delta$ for all $a \in \mathcal{G}$, where $\mathcal{G} \subset C(X)$ is a finite subset and $\delta > 0$. Suppose that $\mathrm{bott}_1(h, u)$ is well defined and $\mathrm{bott}_1(h, u) = 0$ and $[u] = 0$. With sufficiently large $\mathcal{G}$ and sufficiently small δ, we denote ψ the ϵ-$\mathcal{F}_0 \otimes S$-multiplicative contractive completely positive map given by 2.8 which satisfies (e 2.3) for $\phi = h$. From the definition, it means, for any $x \in \beta^{(1)}(K_1(C(X)))$ and any $y \in \beta^{(0)}(K_0(C(X)))(= \mathbb{Z})$,

$$[\psi](x) = 0 \text{ and } [\psi](y) = 0,$$

provided that $\mathcal{F}_0$ is large enough and ϵ is small enough. Let $\mathcal{P} \subset \underline{K}(C(X \times S^1)$ be a finite subset. Then, from 1) above, with sufficiently large $\mathcal{G}$ and sufficiently small δ, we may assume that

$$[\psi]|_{\mathcal{P}} = [\phi_0]|_{\mathcal{P}} \text{ in } KL(C(X \times S^1), A),$$

where $\phi_0 : C(X \times S^1) \to A$ is defined by $\phi_0(f \otimes g) = h(f)g(t_0) \cdot 1$ for all $f \in C(X)$ and $g \in C(S^1)$ and where $t_0 \in S^1$ is a fixed point.

3) Let $Y \subset X \times S^1$ be a compact subset and let $s : C(X \times S^1) \to C(Y)$ be the quotient map. It is known and easy to see that s_{*0} maps $\mathrm{ker}\rho_{C(X \times S^1)}$ onto $\mathrm{ker}\rho_{C(Y)}$ (see Lemma 2.2 of [**13**]). In particular, if $h : C(X \times S^1) \to A$ is a homomorphism such that $(h_{*0})|_{\mathrm{ker}\rho_{C(X \times S^1)}} = 0$ then $(h_1)_{*0}|_{\mathrm{ker}\rho_{C(Y)}} = 0$, where $h_1 : C(Y) \to A$ is a homomorphism defined by $h = h_1 \circ s$.

(4) If Y is a compact subset of $X \times S^1$, then $K_i(C(Y))$ is also torsion free. Let $\phi : C(Y) \to A$ be a unital homomorphism for some unital C^*-algebra A. Then $[\phi] = [\phi_{00}]$ for some $\phi_{00} : C(Y) \to A$ which is a point-evaluation, if and only if

$$\phi_{*1} = 0 \text{ and } \phi_{*0}(s_*(\beta_1(K_1(C(X))))) = 0.$$

Recall that a unital C^*-algebra A is said to be K_1-simple, if $K_1(A) = U(A)/U_0(A)$ and if $[p] = [q]$ in $K_0(A)$ for two projections p and q in A, then there exists $w \in A$ such that $w^*w = p$ and $ww^* = q$.

Lemma 3.2. *Let X be a finite CW complex with torsion free $K_1(C(X))$ and let B be a unital K_1-simple C^*-algebra of real rank zero. Suppose that $\lambda : K_1(C(X)) \to K_1(B)$ is a homomorphism. Then, for any $\delta > 0$, any finite subset $\mathcal{G} \subset C(X)$, any finite subset $\mathcal{P}_1 \subset K_1(C(X))$ and any finite subset $\mathcal{P}_0 \subset \mathrm{ker}\rho_{C(X)}$, there exists a δ-$\mathcal{G}$-multiplicative contractive completely positive linear map $L : C(X) \to B$ such that*

$$[L]|_{\mathcal{P}_1} = \lambda|_{\mathcal{P}_1} \text{ and } [L]|_{\mathcal{P}_0} = 0.$$

If, in addition, X has dimension 1, then there is a unital monomorphism $h : C(X) \to B$ such that

$$h_{1*} = 0.$$

PROOF. If B is a finite dimensional C^*-algebra we define $L(f) = f(\xi)1_B$ for all $f \in C(X)$, where $\xi \in X$ is a fixed point. We now assume that B is non-elementary. It is also clear that one may reduce the general case to the case that X is connected.

Write $K_1(C(X)) = \mathbb{Z}^k$ for some integer $k > 0$. Let C be a unital simple A$\mathbb{T}$-algebra of real rank zero with $K_1(C) = \mathbb{Z}^k$ and $\ker\rho_C = \{0\}$. For convenience, one may assume that $C = \overline{\cup_{n=1}^{\infty} C_n}$, where

$$C_n = \bigoplus_{i=1}^{k} M_{r(i)}(C(S^1)), n = 1, 2....$$

The construction of such A$\mathbb{T}$-algebra is standard.

It follows from a result of [**20**] that there is a unital homomorphism $h_0 : C(X) \to C$ such that

$$(\text{e}\,3.1) \qquad (h_0)_{*1} = \text{id}_{\mathbb{Z}^k} \text{ and } ((h_0)_{*0})|_{\ker\rho_{C(X)}} = 0.$$

(see also [**26**]). For any finite subset $\mathcal{G} \subset C(X)$, we may assume that $h_0(\mathcal{G}) \subset C_n$ for $n \geq 1$. Nuclearity of C^*-algebras involved implies that there is a δ-$\mathcal{G}$-multiplicative contractive completely positive linear map $L' : C(X) \to C_n$ such that

$$\|h_0(f) - L'(f)\| < \delta \text{ for all } \mathcal{G}.$$

We may also assume (by choosing a larger n) that $\overline{\mathcal{Q}_{\delta,\mathcal{G}}} \supset \mathcal{P}_0 \cup \mathcal{P}_1$ and that

$$[L']|_{\mathcal{P}_1} = \text{id}_{\mathbb{Z}^k}|_{\mathcal{P}_1} \text{ and } [L']|_{\mathcal{P}_0} = 0.$$

There are non-zero mutually orthogonal projections $e_{j,i} \in B$ ($j = 1, 2, ..., r(i)$, $i = 1, 2, .., k$) such that for fixed k, $\{e_{j,k} : i = 1, 2, ..., r(i)\}$ is a set of non-zero mutually orthogonal projections.

Let $\{g_i : i = 1, 2, ..., k\}$ be the standard generators for $\mathbb{Z}^k$ and let $z_i = \lambda(g_i)$, $i = 1, 2, ..., k$. There is a unitary $v_i \in e_{1,i}Be_{1,i}$ such that $[v_i + (1 - e_{1,i})] = z_i$, $i = 1, 2, ...$ Put $p = \sum_{i=1}^{k} \sum_{j=1}^{r(i)} e_{i,j}$. We then obtain a unital monomorphism $h_1 : C_n \to pBp$ such that

$$(h_1)_{*1} = \lambda$$

is a homomorphism from $\mathbb{Z}^k$ to $K_1(B)$. Define $h_{00} : C(X) \to (1-p)A(1-p)$ by $h_{00}(f) = f(\xi)(1-p)$ for all $f \in C(X)$, where $\xi \in X$ is a fixed point in X. Define $L = (h_1 \circ L' \circ h_0) \oplus h_{00}$. where $\xi \in X$ is a fixed point in X. It is clear that so defined L meets the requirements.

By 2.25, there is a unital monomorphism $h_{00}' : C(X) \to (1-p)B(1-p)$ such that $[h_{00}'] = [h_{00}]$ in $KK(C(X), B)$. The last statement follows from the fact that $C(X)$ is semi-projective (by 5.1 of [**42**]) and replacing $h_1 \circ L' \circ h_0$ by a homomorphism and replacing h_{00} by h_{00}'. □

Lemma 3.3. *Let X be a connected finite CW complex with dimension 1 (with a fixed metric), let $A \in \mathbf{B}$ and let $h : C(X) \to A$ be a monomorphism. Suppose that there is a unitary $u \in A$ such that*

$$(\text{e}\,3.2) \qquad u = uh(a) \textit{ for all } a \in C(X), \ [u] = 0 \textit{ in } K_1(A) \textit{ and } \text{bott}_1(h, u) = 0.$$

Suppose also that $\psi : C(X \times S^1) \to A$ defined by $\psi(f \otimes g) = h(f)g(u)$ for $f \in C(X)$ and $g \in C(S^1)$ is a monomorphism. Then, for any $\epsilon > 0$ and any finite subset $\mathcal{F} \subset C(X)$, there exists a rectifiable continuous path of unitaries $\{u_t : t \in [0, 1]\}$ of A such that

$$(\text{e}\,3.3) \qquad u_0 = u, \ u_1 = 1_A \textit{ and } \|[h(a), u_t]\| < \epsilon$$

for all $a \in \mathcal{F}$ and all $t \in [0,1]$. Moreover,

$$\text{Length}(\{u_t\}) \le \pi + \epsilon\pi. \tag{e 3.4}$$

PROOF. Let $\epsilon > 0$ and $\mathcal{F} \subset C(X)$ be a finite subset. We may assume that $1_{C(X)} \in \mathcal{F}$. Let

$$\mathcal{F}' = \{f \otimes a : f \in \mathcal{F}, a = 1_{C(S^1)}, \text{or } \ a = z\}.$$

Let $\sigma_1 = \sigma_{X\times S^1, \mathcal{F}', \epsilon/32}$ (see 2.19). Let $\delta_1 > 0$, $\mathcal{G} \subset C(X \times S^1)$ and $\mathcal{P} \subset \underline{K}(C(X \times S^1))$ be required by Theorem 1.12 of [25] associated with $X \times S^1$, $\mathcal{F}'$, $\epsilon/16$ (in place of ϵ). Here we also assume that any two δ_1-$\mathcal{G}$-multiplicative contractive completely positive linear maps L_1, L_2 from $C(X \times S^1)$ satisfy

$$[L_1]|_{\mathcal{P}} = [L_2]|_{\mathcal{P}}, \ \text{if} \tag{e 3.5}$$

$$L_1 \approx_{\delta_1} L_2 \ \text{ on } \mathcal{G}. \tag{e 3.6}$$

Without loss of generality, we may assume that $\mathcal{G} = \mathcal{G}_1 \otimes S$, where $\mathcal{G}_1$ is in the unit ball of $C(X)$ and $S = \{1_{C(S^1)}, z\}$. We may also assume that $\delta_1 < \epsilon/2$ and that $\mathcal{F} \subset \mathcal{G}_1$.

Let $\eta_1 = (1/2)\sigma_{X\times S^1, \mathcal{G}, \min\{\epsilon/32, \delta_1/32\}}$. So if $\text{dist}(x, x') < \eta_1$ and if $\text{dist}(t, t') < \eta_1$, then

$$|f(x \times t) - f(x' \times t')| < \min\{\epsilon/32, \delta_1/32\} \ \text{ for all } \ f \in \mathcal{G}. \tag{e 3.7}$$

Let $\{x_1, x_2, ..., x_m\}$ be $\sigma_1/4$-dense in X and let $\{t_1, t_2, ..., t_l\}$ divide the unit circle into l arcs with the length with $2\pi/l < \eta_1$. We may assume that $\eta_1 < \sigma_1/4$. We also assume that $t_1 = 1$. In particular, $\{x_i \times t_j : i = 1, 2, ..., m, j = 1, 2, ..., l\}$ is $\sigma_1/2$-dense.

Let $g_{i,j}$ be nonnegative functions in $C(X \times S^1)$ with $0 \le g_{i,j} \le 1$ such that $g_{i,j}(\xi) = 1$ if $\text{dist}(\xi, x_i \times t_j) < \eta_1/4s$ and $g_{i,j}(\xi) = 0$ if $\text{dist}(\xi, x_i \times t_j) \ge \eta_1/2s$. Since A is a unital simple C^*-algebra with real rank zero and ψ is injective, there are non-zero, mutually orthogonal and mutually equivalent projections $E_{i,j}, E'_{i,j} \in \overline{\psi(g_{i,j})A\psi(g_{i,j})}$ for each i and j. Since A is simple and has real rank zero, then (by repeated application of (1) of Lemma 3.5.6 of [29], for example) there is a non-zero projection $E_0 \le E_{1,1}$ such that

$$[E_0] \le [E_{i,j}], \ i = 1, 2, ..., m \ \text{ and } \ j = 1, 2, ..., l. \tag{e 3.8}$$

Put

$$\Phi(f) = \sum_{i,j} f(x_i \times t_j)(E_{i,j} + E'_{i,j}) + \tilde{\phi}(f) \ \text{ for all } \ f \in C(X \times S^1), \tag{e 3.9}$$

where

$$\tilde{\phi}(f) = (1 - \sum_{i,j}(E_{i,j} + E'_{i,j}))\psi(f)(1 - \sum_{i,j}(E_{i,j} + E'_{i,j})) \ \text{ and} \tag{e 3.10}$$

$$\Phi_1(f) = f(x_1 \times t_1)(E'_{1,1} + E_{1,1} - E_0) + \sum_{(i,j)\neq(1,1)} f(x_i \times t_j)(E_{i,j} + E'_{i,j}) +$$

$$+\tilde{\phi}(f) \tag{e 3.11}$$

for $f \in C(X \times S^1)$. Thus, by the choice of σ_1,

$$\|\psi(f) - \Phi(f)\| < \epsilon/32 \ \text{ for all } \ f \in \mathcal{G}. \tag{e 3.12}$$

By the choice of η_1, one sees that $\tilde{\phi}$ is δ_1-$\mathcal{G}$-multiplicative. So Φ_1 is δ_1-$\mathcal{G}$-multiplicative. Moreover, $\Sigma_{\delta_1}(\Phi_1, \mathcal{F})$ is $\sigma_1/2$-dense, since $\{(x_i \times t_j) : i, j\}$ is $\sigma_1/2$-dense.

There are two non-zero mutually orthogonal projections $e_1, e_2 \in E_0AE_0$ such that $e_1 + e_2 = E_0$. Since both e_1Ae_1 and e_2Ae_2 are simple C^*-algebras in $\mathbf{B}$, and X is a connected finite CW complex, it is easy to construct unital monomorphisms $h_1 : C(X) \to e_1Ae_1$ and $h_2 : C(X) \to e_2Ae_2$ such that

$$(h_1)_{*1} = h_{*1} \text{ and } (h_2)_{*1} + h_{*1} = 0 \tag{e 3.13}$$

(by 3.2). It follows from Cor.1.14 of [**25**] that there are mutually orthogonal projections $\{p_1, p_2, ..., p_K\} \subset E_0AE_0$ and points $\{y_1, y_2, ..., y_K\} \subset X$ such that $\sum_{i=1}^K p_i = E_0$ and

$$\|(h_1(f) + h_2(f)) - \sum_{k=1}^K f(y_k)p_k\| < \epsilon/32 \text{ for all } f \in \mathcal{G}_1. \tag{e 3.14}$$

Define $\psi_1(f \otimes g) = h_1(f)g(1)e_1$ and $\psi_2(f \otimes g) = h_2(f)g(1)e_2$ for all $f \in C(X)$ and $g \in C(S^1)$. Define $\psi_0 : C(X \times S^1) \to E_0AE_0$ by

$$\psi_0(f \otimes g) = \sum_{k=1}^K f(y_k)g(1)p_k \tag{e 3.15}$$

for all $f \in C(X)$ and $g \in C(S^1)$. By replacing $\epsilon/32$ by $\epsilon/16$, in (e 3.14), we may assume that $\{y_1, y_2, ..., y_K\} \subset \{x_1, x_2, ..., x_{K'}\}$ with $K' \le m$. Keeping this in mind, since $p_k \le e_2$, by (e 3.8), we obtain a unitary $w_1 \in A$ such that

$$\mathrm{ad}\, w_1 \circ (\psi_0(f) + \Phi_1(f)) = \Phi(f) \tag{e 3.16}$$

for all $f \in C(X \times S^1)$. It follows that (by (e 3.14)

$$\mathrm{ad}\, w_1 \circ ((\psi_1 \oplus \psi_2) + \Phi_1) \approx_{\epsilon/16} \Phi \text{ on } \mathcal{G} \tag{e 3.17}$$

and (by e 3.12))

$$\mathrm{ad}\, w_1 \circ ((\psi_1 \oplus \psi_2) + \Phi_1) \approx_{3\epsilon/32} \psi \text{ on } \mathcal{G}. \tag{e 3.18}$$

It follows from (e 3.13) and the condition that $\mathrm{bott}_1(h, u) = 0$ and $[u] = 0$ that

$$[\mathrm{ad}\, w_1 \circ (\psi_2 \oplus \Phi_1)]|_{\mathcal{P}} = [\Phi_{00}]|_{\mathcal{P}}, \tag{e 3.19}$$

where $\Phi_{00} = f(\xi_X \times 1)(1 - e_1)$ for all $f \in C(X \times S^1)$. We also check that $\mathrm{ad}\, w_1 \circ (\psi_2 \oplus \Phi_1)$ is δ_1-$\mathcal{G}$-multiplicative and $\Sigma_{\delta_1}(w_1 \circ (\psi_2 \oplus \Phi_1), \mathcal{F}')$ is $\sigma_1/2$-dense. It follows from Theorem 1.12 of [**25**] that there is a homomorphism $\Phi_0 : C(X \times S^1) \to (1 - e_1)A(1 - e_1)$ with finite dimensional range such that

$$\mathrm{ad}\, w_1 \circ (\psi_2 \oplus \Phi_1) \approx_{\epsilon/16} \Phi_0 \text{ on } \mathcal{F}'. \tag{e 3.20}$$

In the finite dimensional commutative C^*-subalgebra $\Phi_0(C(X \times S^1)$, there is a continuous rectifiable path of unitaries $\{U_t : t \in [0, 1]\}$ of $(1 - e)A(1 - e)$ such that

$$U_0 = \Phi_0(1 \otimes z), U_1 = 1 - w_1^* e_1 w_1 \text{ and } \Phi_0(f \otimes 1)U_t = U_t\Phi_0(f \otimes 1) \tag{e 3.21}$$

for all $t \in [0, 1]$ and $f \in C(X)$. Moreover

$$\mathrm{Length}(\{U_t\}) \le \pi. \tag{e 3.22}$$

Define $v_t = w_1^* e_1 w_1 \oplus U_t$ for $t \in [0,1]$. Then, (note that $\psi_1(1 \otimes z) = e_1$) by (e 3.20) and (e 3.18),

$$\begin{aligned} \|v_0 - u\| &= \|w_1^* e_1 w_1 \oplus \Phi_0(1 \otimes z) - u\| \\ &\le \|w_1^* e_1 w_1 \oplus [\Phi_0(1 \otimes z) - w_1^*(\psi_2 \oplus \Phi_1)(1 \otimes z) w_1]\| \\ &+ \|w_1^* e_1 w_1 \oplus w_1^*(\psi_2 \oplus \Phi_1)(1 \otimes z) w_1 - u\| \\ &< \epsilon/16 + \|\mathrm{ad}\, w_1^*((\psi_1 \oplus \psi_2)(1 \otimes z) \oplus \Phi_1(1 \otimes z) w_1 - \psi(1 \otimes z)\| \\ &< \epsilon/16 + 3\epsilon/32 = \epsilon/8. \end{aligned} \tag{e 3.23}$$

We estimate that, by (e 3.18) and (e 3.20) and (e 3.21),

$$\|[h(f), v_t]\| < 6\epsilon/32 + \|[\mathrm{ad}\, w_1 \circ ((\psi_1 + \psi_2) + \Phi_1)(f \otimes 1), v_t]\| \tag{e 3.24}$$

$$< 3\epsilon/16 + \epsilon/8 + \|[\mathrm{ad}\, w_1 \circ (\psi_1 + \Phi_0)(f \otimes 1), v_t]\| \tag{e 3.25}$$

$$= 5\epsilon/16 \tag{e 3.26}$$

for all $f \in \mathcal{F}$. Combing this with (e 3.23), we obtain a continuous rectifiable path of unitaries $\{u_t : t \in [0,1]\}$ of A such that

$$u_0 = u, \quad u_1 = 1_A \text{ and } \|[h(a), u_t]\| < \epsilon \text{ for all } a \in \mathcal{F} \tag{e 3.27}$$

and all $t \in [0,1]$. Moreover,

$$\mathrm{Length}(\{u_t\}) \le \pi + \epsilon\pi. \tag{e 3.28}$$

□

Lemma 3.4. *Let X_1 be a connected finite CW complex with dimension 1. Let $\epsilon > 0$ and $\mathcal{F}_1 \subset C(X_1)$ be a finite subset. There is $\sigma > 0$ satisfying the following: Suppose that $A \in \mathbf{B}$ is an infinite dimensional unital C^*-algebra, suppose that $h_1 : C(X_1) \to A$ is a homomorphism and suppose that $u \in A$ is a unitary with $[u] = 0$ in $K_1(A)$ such that*

$$h_1(a)u = uh_1(a) \text{ for all } a \in C(X_1) \text{ and } \mathrm{bott}_1(h_1, u) = 0, \tag{e 3.29}$$

and suppose that X is a subset of X_1 which is a finite CW complex and the spectrum of h_1 is $Y \subset X$ which is σ-dense in X. Then there exists a unital monomorphism $\bar{h} : C(X \times S^1) \to A$ and a rectifiable continuous path of unitaries $\{u_t : t \in [0,1]\}$ in A such that

$$u_0 = u, \quad u_1 = \bar{h}(1 \otimes z), \quad \|[h_1(a), u_t]\| < \epsilon \text{ and} \tag{e 3.30}$$

$$\|\bar{h}(s(a) \otimes 1) - h_1(a)\| < \epsilon \text{ for all } a \in \mathcal{F}_1, \tag{e 3.31}$$

where $s : C(X_1) \to C(X)$ is defined by $s(f) = f|_X$. Moreover

$$\mathrm{Length}(\{u_t\}) \le \pi + \epsilon\pi \tag{e 3.32}$$

Proof. There is a homomorphism $h : C(X) \to A$ such that $h = h_1 \circ s$, where $s : C(X_1) \to C(X)$ is the quotient map. Put $\mathcal{F}_1 = s(\mathcal{F})$ and

$$\mathcal{F}' = \{f \otimes g : f \in \mathcal{F}_1 \text{ and } g = 1, \text{ or } g = z\}.$$

Define $\psi : C(X \times S^1) \to A$ by $\psi(f \otimes g) = h(f)g(u)$ for $f \in C(X)$ and $g \in C(S^1)$. By (e 3.29), $[\psi]|_{\ker \rho_{C(X \times S^1)}} = 0$.

Large part of the argument is the same as that used in 3.3. So we will keep the notation there and keep all the proof from the beginning of that proof to the equation (e 3.7). Moreover, we will refer to Theorem 2.5 of [**25**] instead of Lemma 1.12 of [**25**] and δ_1, $\mathcal{G}$ and $\mathcal{P}$ are now stated in Theorem 2.5 of [**25**] (for $X \times S^1$).

Note that we still assume that $\mathcal{F} \subset \mathcal{G}_1$. Note that we have $\overline{\mathcal{Q}_{\delta_1,\mathcal{G}}} \subset \mathcal{P}$ (see 2.3). It should be noted (see 1.7 of [**25**]) that

$$\sigma_{X_1 \times S^1, \mathcal{F}\otimes S, \epsilon/32} \le \sigma_{X \times S^1, \mathcal{F}', \epsilon/32} = \sigma_1,$$

where $S = \{1, z\} \subset C(S^1)$. Put $\sigma = \sigma_{X_1 \times S^1, \mathcal{F}\otimes S, \epsilon/32}/4$. Let Y be σ-dense in X. We assume that $\{x_1, x_2, ..., x_m\} \subset Y$ is σ -dense in X.

Without loss of generality, we may still assume, for each i, that there is j such that $\psi(g_{i,j}) \neq 0$. We may replace $\{t_1, t_2, ..., t_l\}$ by $\{t_{i(1)}, t_{i(2)}, ...t_{i(k(i))}\}$ with $t_{i(1)} = 1$. We assume that $\{x_i \times t_{i,j} : j = 1, 2, ..., k(i), i = 1, 2, ..., m\}$ is σ -dense in Y. We now replace $g_{i,j}$ by $g_{i,1}$, $E_{i,j}$ by $E_{i,1}$ and $E'_{i,j}$ by $E'_{i,1}$. Accordingly, (e 3.8), (e 3.9),(e 3.10),(e 3.11) and (e 3.12) hold where we replace all j by 1, t_1 by $t_{1(1)}$ and t_j by $t_{i(1)} = 1$, respectively.

Now, instead of taking two projections in E_0AE_0, we choose $e_1 = E_0$. We also introduce another homomorphism F_3 as follows. By 2.25, there are monomorphisms $F_1 : C(X \times S^1) \to C(D)$ and $F_2 : C(D) \to e_1Ae_1$, where D is a finite union of closed intervals. Put $F_3 = F_2 \circ F_1$. It follows from Cor. 1.14 of [**25**] that there is a unital homomorphism $\psi_0 : C(X \times S^1) \to e_1Ae_1$ with finite dimensional range such that

$$\|F_3(f) - \psi_0(f)\| < \min\{\epsilon/32, \delta_1/32\} \tag{e 3.33}$$

for all $f \in \mathcal{G}$. Without loss of generality, by changing $\epsilon/32$ to $2\epsilon/32 = \epsilon/16$, we may assume that

$$\psi_0(f \otimes g) = \sum_{i,j} f(x_i)g(t_j)p_{i,j} \tag{e 3.34}$$

for all $f \in C(X)$ and $g \in C(S^1)$ and $\{p_{i,j}\}$ is a set of finitely many mutually orthogonal projections in e_1Ae_1 with $\sum_{i,j} p_{i,j} = e_1$. Define another homomorphism $\psi_{00} : C(X \times S^1) \to e_1Ae_1$ such that

$$\psi_{00}(f \otimes g) = \sum_{i=1}^{m} f(x_i)g(1)q_i \tag{e 3.35}$$

for all $f \in C(X)$ and $g \in C(S^1)$, where $q_i = \sum_j p_{i,j}$, $i = 1, 2, ..., m$. Working in the finite dimensional commutative C^*-subalgebra $\psi_0(C(X \otimes S^1))$, it is easy to obtain a continuous rectifiable path of unitaries $\{v_t : t \in [0, 1]\}$ of e_1Ae_1 such that

$$v_0 = \psi_{00}(1 \otimes z), \;\; v_1 = \psi_0(1 \otimes z) \;\text{ and }\; v_t\psi_{00}(f) = \psi_{00}(f)v_t \tag{e 3.36}$$

for all $f \in C(X \times S^1)$ and $t \in [0, 1]$. Moreover,

$$\text{Length}(\{v_t\}) \le \pi. \tag{e 3.37}$$

Since $q_i \le e_1 = E_0$, the same argument in the proof of 3.3 provides a unitary $w_1 \in A$ such that

$$\text{ad}\, w_1 \circ (\psi_{00}) + \Phi_1)(f) = \Phi(f) \tag{e 3.38}$$

for all $f \in C(X \times S^1)$. It follows, as in the proof of 3.3, that

$$\text{ad}\, w_1 \circ (\psi_{00} + \Phi_1)(f) \approx_{\epsilon/16} \psi(f) \;\text{ on }\; f \in \mathcal{G} \tag{e 3.39}$$

Put $V_t = w_1^*[v_t \oplus \Phi_1(1 \otimes z)]w_1$ for $t \in [0, 1]$. Then by (e 3.36) and (e 3.39),

$$\|[V_t, \psi(f \otimes 1)]\| < \epsilon/8 \tag{e 3.40}$$

for all $f \in \mathcal{F}$. We have, by (e 3.39),

$$\|V_0 - u\| = \|w_1^*(v_0 \oplus \Phi_1(1 \otimes z))w_1 - \psi(1 \otimes z)\| < \epsilon/8 \tag{e 3.41}$$

Note that

$$[\mathrm{ad} w_1 \circ (\psi_{00}) + \Phi_1]|_{\mathcal{P}} = [\psi]|_{\mathcal{P}}.$$

Since ψ_{00} has finite dimensional range and $[\psi]|_{\ker \rho_{C(X)}} = 0$, by applying Theorem 2.5 of [**25**], one obtains a homomorphism $\Phi_0 : C(X \times S^1) \to (1 - E_0)A(1 - E_0)$ such that

$$\Phi_1 \approx_{\epsilon/16} \Phi_0 \ \text{ on } \mathcal{F}'. \tag{e 3.42}$$

Now define $\bar{h} : C(X \times S^1) \to A$ by

$$\bar{h}(f) = \mathrm{ad}\, w_1 \circ (F_3 \oplus \Phi_0)(f) \ \text{ for all } \ f \in C(X \times S^1). \tag{e 3.43}$$

By (e 3.42) and (e 3.39), we have

$$\|\bar{h}(f \otimes 1) - h(f)\| < \epsilon/8 \ \text{for all} \ f \in \mathcal{F}. \tag{e 3.44}$$

We estimate that, by (e 3.33) and (e 3.42),

$$\|V_1 - \bar{h}(1 \otimes z)\| = \|w_1^*(\psi_0(1 \otimes z) \oplus \Phi_1(1 \otimes z))w_1 - \bar{h}(1 \otimes z)\| \tag{e 3.45}$$

$$< \epsilon/32 + \epsilon/16 \tag{e 3.46}$$

Note that $u = \psi(1 \otimes z)$. By connecting V_1 with $\bar{h}(1 \otimes z)$ and V_0 to u appropriately, we obtain a continuous rectifiable path of unitaries $\{u_t : t \in [0,1]\}$ in A such that

$$u_0 = u, \ \ u_1 = \bar{h}(1 \otimes z) \ \text{ and } \ \|[u_t, h(f)]\| < \epsilon \tag{e 3.47}$$

for all $f \in \mathcal{F}$ and $t \in [0,1]$. Moreover,

$$\mathrm{Length}(\{u_t\}) \le \pi + \epsilon\pi. \tag{e 3.48}$$

Since F_3 is a monomorphism, we conclude that $\bar{h}$ is also a monomorphism. □

Remark 3.5. In the statement of 3.4, define $h_2 : C(X_1) \to A$ by $h_2(f) = \bar{h}(s(f) \otimes 1)$ for all $f \in C(X_1)$. Note that, if ϵ is small enough and $\mathcal{F}$ is large enough, then $\mathrm{bott}_1(h_2, u_1) = \mathrm{bott}_1(h, u) = 0$ (see 2.7). Since X has dimension 1, $s_{*1}(K_1(C(X_1))) = K_1(C(X))$. This implies that $\mathrm{bott}_1(h_2', u_1) = 0$, where $h_2' : C(X) \to A$ by $h_2'(f) = \bar{h}(f \otimes 1)$ for $f \in C(X)$.

It should be noted that in the statement of 3.3 and 3.4, the length of $\{u_t\}$ can be controlled by $\pi + \epsilon$.

Lemma 3.6. *Let X be a connected finite CW complex with dimension 1. Then, for any $\epsilon > 0$ and any finite subset $\mathcal{F} \subset C(X)$, there is $\sigma > 0$ satisfying the following: Suppose that $A \in \mathbf{B}$ is not finite dimensional and that $h : C(X) \to A$ is a unital homomorphism whose spectrum is σ-dense in X and suppose that there is a unitary $u \in A$ with $[u] = 0$ in $K_1(A)$ such that*

$$h(a)u = uh(a) \ \textit{ for all } \ a \in C(X) \ \textit{ and } \ \mathrm{bott}_1(h, u) = 0. \tag{e 3.49}$$

Then, there exists a rectifiable continuous path of unitaries $\{u_t : t \in [0,1]\}$ of A such that

$$u_0 = u, \ \ u_1 = 1_A \ \textit{ and } \ \|[h(a), u_t]\| < \epsilon \tag{e 3.50}$$

for all $a \in \mathcal{F}$ and all $t \in [0,1]$. Moreover,

$$\mathrm{Length}(\{u_t\}) \le 2\pi + \epsilon\pi. \tag{e 3.51}$$

PROOF. Let $\epsilon > 0$ and $\mathcal{F} \subset C(X)$ be a finite subset. We may assume that $\mathcal{F}$ is in the unit ball of $C(X)$. Let $\mathcal{P} \subset K_1(C(X))$ be a finite subset containing a set of generators. Let $\epsilon_1 > 0$ and $\mathcal{F}_1 \subset C(X)$ be a finite subset so that $\|[h'(a), w]\| < \epsilon_1$ for all $a \in \mathcal{F}_1$ implies that $\text{bott}_1(h, w)$ is well defined for any unital homomorphism $h' : C(X) \to A$ and any unitary $w \in A$. We may assume that $\epsilon_1 < \epsilon$ and $\mathcal{F} \subset \mathcal{F}_1$.

Let σ be in 3.4 corresponding to $\epsilon_1/4$ and $\mathcal{F}_1$. There is a subset $Y \subset X$ which is a finite CW complex so that the spectrum of h is σ-dense in Y. Then it is clear that the lemma follows from 3.3 and 3.4 (see also 3.5). □

Theorem 3.7. *Let X be a finite CW complex with dimension 1. Then, for any $\epsilon > 0$ and any finite subset $\mathcal{F} \subset C(X)$, there exists $\delta > 0$, a finite subset $\mathcal{G} \subset C(X)$ and $\sigma > 0$ satisfying the following:*

Suppose that $A \in \mathbf{B}$, suppose that $h : C(X) \to A$ is a unital homomorphism whose spectrum is σ-dense in X and suppose that there is a unitary $u \in A$ such that

$$\|[h(a), u]\| < \delta \text{ for all } a \in \mathcal{G}, \ \text{bott}_0(h, u) = 0 \text{ and } \text{bott}_1(h, u) = 0. \tag{e 3.52}$$

Then there exists a rectifiable continuous path of unitaries $\{u_t : t \in [0, 1]\}$ of A such that

$$u_0 = u, \ u_1 = 1_A \text{ and } \|[h(a), u_t]\| < \epsilon \tag{e 3.53}$$

for all $a \in \mathcal{F}$ and all $t \in [0, 1]$. Moreover,

$$\text{Length}(\{u_t\}) \le 2\pi + \epsilon. \tag{e 3.54}$$

PROOF. We assume that A is not finite dimensional. So A is not elementary. The case that A is finite dimensional will be dealt with next in 3.11. By considering each connected component of X, since we assume that $\text{bott}_0(h, u) = 0$, one easily reduces the general case to the case that X is connected (see also the beginning of the proof of 7.4). So for the rest of this proof we assume that X is connected and $[u] = 0$ in $K_1(A)$. Fix $\epsilon > 0$ and a finite subset $\mathcal{F} \subset C(X)$. We may assume that $1 \in \mathcal{F}$. Let $\mathcal{F}' = \{f \otimes g : f \in \mathcal{F} \text{ and } g = 1, \text{or } g = z\}$. Let $\sigma_1 > 0$ (in place of σ) be in 3.6 corresponding to $\epsilon/2$ and $\mathcal{F}$. There exists $\eta_0 > 0$ and a finite subset $\mathcal{G}_0 \subset C(X)$ satisfy the following: if $h_1, h_2 : C(X) \to B$ (for some unital C^*-algebra B) are unital homomorphisms such that

$$\|h_1(f) - h_2(f)\| < \eta_0 \text{ for all } f \in \mathcal{G}_0,$$

then the spectrum of h_2 is σ_1-dense in X, provided that the spectrum of h_1 is $\sigma_1/2$-dense.

Put $\eta = \min\{\eta_0/2, \epsilon/2\}$ and $\sigma = \sigma_1/2$. Let $\delta_1 > 0$, $\mathcal{G}_1 \subset C(X \times S^1)$ and $\mathcal{P} \subset \underline{K}(C(X \times S^1))$ be a finite subset required by Theorem 2.5 of [**25**] corresponding to $\eta/2$ and $\mathcal{F}'$. We may assume that $\mathcal{P}$ contains a set of generators of $\beta_1(K_1(C(X)))$ and $\delta_1 < \eta$. We may assume that δ_1 is sufficiently small and $\mathcal{G}_2$ is sufficiently large so that for any two δ_1-$\mathcal{G}_2$-multiplicative contractive completely positive linear maps $L_i : C(X \times S^1) \to B$ (for any unital C^*-algebra), $[L_i]|_{\mathcal{P}}$ is well-defined and

$$[L_1]|_{\mathcal{P}} = [L_2]|_{\mathcal{P}}, \tag{e 3.55}$$

provided that

$$\|L_1(f) - L_2(f)\| < \delta_1 \text{ for all } f \in \mathcal{G}_1.$$

Moreover, by choosing even smaller δ_1, we may assume that $\mathcal{G}_1 = \mathcal{G}' \otimes S$, where $\mathcal{G}' \subset C(X)$ and $S = \{1_{C(S^1)}, z\}$. We may also assume that $\mathcal{G}_0 \subset \mathcal{G}'$.

Let $\delta > 0$ and $\mathcal{G} \subset C(X)$ (in place of $\mathcal{F}_1$) be a finite subset required in 2.8 for δ_1 (in place of ϵ) and $\mathcal{G}'$ (in place of $\mathcal{F}$).

Now suppose that h and u satisfy the conditions in the theorem for the above δ, $\mathcal{G}$ and σ. Let $\psi : C(X \times S^1) \to A$ be given by 2.8 for $\phi = h$. The condition that $\mathrm{bott}_1(h, u) = 0$, implies that

$$[\psi]|_{\mathcal{P}} = \alpha(\mathcal{P}) \tag{e 3.56}$$

for some $\alpha \in \mathcal{N}k$ (see 2.1 of [**25**]). By 2.8 and Theorem 2.5 of [**25**], there is a unital homomorphism $H : C(X \times S^1) \to A$ such that

$$\|H(f \otimes g) - \psi(f \otimes g)\| < \eta_1/2 \text{ for all } f \in \mathcal{G}_1 \text{ and } g = 1, \text{ or } g = z. \tag{e 3.57}$$

Put $h_1(f) = H(f \otimes 1)$ (for $f \in C(X)$ and $v = H(1 \otimes z)$. Then

$$[h_1, v] = 0, \ \ \mathrm{bott}_1(h_1, v) = 0 \ \text{ and } \ [v] = 0 \ \text{ in } \ K_1(A).$$

Note that, by the choice of η_1, the spectrum of h_1 is σ_1-dense in X. Thus 3.6 applies to h_1 and v (with $\epsilon/2$). By (e 3.57),

$$h_1 \approx_{\eta_1/2} h \text{ on } \mathcal{G}_1. \tag{e 3.58}$$

The lemma follows. □

Remark 3.8. Note that, in the proof of 3.4, σ in the statement of 3.4 is chosen to be $\sigma_{X_1 \times S^1, \mathcal{F} \otimes S, \epsilon/32}$ (see 2.19). Thus, from the proof of 3.6, σ in 3.6 can be $\sigma_{X \times S^1, \mathcal{F}_1 \otimes S, \epsilon_1}$ for $0 < \epsilon_1 < \epsilon/32$ and for some finite subset $\mathcal{F}_1 \supset \mathcal{F}$. Here the requirement for ϵ_1 and $\mathcal{F}_1$ is that $\mathrm{bott}_1(h', w)$ is well defined as long as $\|[h'(f), w]\| < \epsilon_1$ for all $f \in \mathcal{F}_1$, where $h' : C(X) \to A$ is any unital homomorphism and $w \in A$ is a unitary. The proof of 3.7 shows that σ in 3.7 can be chosen to be $(1/2)(\sigma_{X \times S^1, \mathcal{F}_1 \otimes S, \epsilon_1})$. If we choose the metric on $X \times S^1$ as in 2.14, then

$$\sigma_{X, \mathcal{F}_1, \epsilon_1/4(M+1)} \le \sigma_{X \times S^1, \mathcal{F}_1 \otimes S, \epsilon_1},$$

where $M = \max\{\|f\| : f \in \mathcal{F}_1\}$.

It is important to note that δ and $\mathcal{G}$ do not depend on σ. Suppose that $e_1, e_2, ..., e_n$ are mutually orthogonal projections in $C(X)$ associated with each component of X. We note that the condition that $\mathrm{bott}_0(h, u) = 0$ in 3.7 means the normal partial isometry close to $h(e_i)uh(e_i)$ gives zero element in $K_1(A)$. This condition is used to reduce the general case to the case that X is connected. If $Y \subset C(X)$ is a compact subset of X, then Y may have more components. Suppose that the spectrum of h is Y. Let $\bar{h} : C(Y) \to A$ be the unital monomorphism induced by h. Then a version of 3.7 will hold if the condition $\mathrm{bott}_0(\bar{h}, u) = 0$ is added.

The following statement does not mention $\mathrm{bott}_0(h, u)$ since we assume that X is connected.

Corollary 3.9. *Let X be a connected finite CW complex with dimension no more than 1. Then, for any $\epsilon > 0$ and any finite subset $\mathcal{F} \subset C(X)$, there exist $\delta > 0$, a finite subset $\mathcal{G} \subset C(X)$ and $\sigma > 0$ satisfying the following:*

Suppose that A be a unital simple C^*-algebra in **B**, *suppose that $h : C(X) \to A$ is a unital homomorphism whose spectrum is σ-dense in X and suppose that there exists a unitary $u \in A$ with $[u] = 0$ in $K_1(A)$ such that*

$$\|[h(f), u]\| < \delta \text{ for all } f \in \mathcal{G} \text{ and } \mathrm{bott}_1(h, u) = 0. \tag{e 3.59}$$

Then there exists a rectifiable continuous path of unitaries $\{u_t : t \in [0,1]\}$ *of* A *such that*

$$u_0 = u, \quad u_1 = 1_A \text{ and } \|[h(a), u_t]\| < \epsilon \tag{e 3.60}$$

for all $a \in \mathcal{F}$ *and all* $t \in [0,1]$. *Moreover,*

$$\mathrm{Length}(\{u_t\}) \le 2\pi + \epsilon. \tag{e 3.61}$$

Note that, in the following corollary, h is a monomorphism.

Corollary 3.10. *Let* X *be a finite CW complex with dimension 1. Then, for any* $\epsilon > 0$ *and any finite subset* $\mathcal{F} \subset C(X)$, *there exist* $\delta > 0$ *and a finite subset* $\mathcal{G} \subset C(X)$ *satisfying the following:*

Suppose that $A \in \mathbf{B}$, *suppose that* $h : C(X) \to A$ *is a unital monomorphism and suppose that there is a unitary* $u \in A$ *such that*

$$\|[h(a), u]\| < \delta \text{ for all } a \in \mathcal{G}, \ \mathrm{bott}_0(h, u) = 0 \text{ and } \mathrm{bott}_1(h, u) = 0. \tag{e 3.62}$$

Then there exists a rectifiable continuous path of unitaries $\{u_t : t \in [0,1]\}$ *of* A *such that*

$$u_0 = u, \quad u_1 = 1_A \text{ and } \|[h(a), u_t]\| < \epsilon \tag{e 3.63}$$

for all $a \in \mathcal{F}$ *and all* $t \in [0,1]$. *Moreover,*

$$\mathrm{Length}(\{u_t\}) \le 2\pi + \epsilon. \tag{e 3.64}$$

The following lemma, in particular, deals with the case that C^*-algebras are finite dimensional. The other reason to include the following is that, in the case that $K_1(A) = 0$, the upper bound for the length can be made shortest possible.

Theorem 3.11. *Let* X *be a finite CW complex of dimension 1. Then, for any* $\epsilon > 0$ *and any finite subset* $\mathcal{F} \subset C(X)$, *there exists* $\delta > 0$ *and a finite subset* $\mathcal{G} \subset C(X)$ *satisfying the following:*

Let $A \in \mathbf{B}$ *with* $K_1(A) = \{0\}$, *let* $h : C(X) \to A$ *be a unital homomorphism and let* $u \in A$ *be a unitary such that*

$$\|[h(g), u]\| < \delta \text{ for all } f \in \mathcal{G} \text{ and } \mathrm{bott}_1(h, u) = 0.$$

Then there exists a continuous rectifiable path of unitaries $\{u_t : t \in [0,1]\}$ *such that*

$$u_0 = u, \quad u_1 = 1_A \text{ and } \|[h(f), u_t]\| < \epsilon \text{ for all } f \in \mathcal{F}.$$

Moreover,

$$\mathrm{Length}(\{u_t\}) \le \pi + \epsilon.$$

Proof. Let $\epsilon > 0$ and $\mathcal{F} \subset C(X)$ be a finite subset. Let $\mathcal{P}_0 \subset K_0(C(X \times S^1))$ be a finite subset which contains a set of generators of $\ker \rho_{C(X \times S^1)}$. There exists $\delta_0 > 0$ and a finite subset $\mathcal{G}_0 \subset C(X \times S^1)$ such that for any two unital δ_0-$\mathcal{G}_0$-multiplicative contractive completely positive linear maps $L_1, L_2 : C(X \times S^1) \to A$ with

$$\|L_1(f) - L_2(f)\| < \delta_0 \text{ for all } f \in \mathcal{G}_0,$$

one has

$$[L_1]|_{\mathcal{P}_0} = [L_2]|_{\mathcal{P}_0}.$$

Without loss of generality, we may assume that $\mathcal{G}_0 = \mathcal{G}_1 \otimes S$, where $\mathcal{G}_1 \subset C(X)$ is a finite subset and $S = \{1_{C(S^1)}, z\}$. Put $\epsilon_1 = \min\{\epsilon/4, \delta_0/2\}$ and $\mathcal{F}_1 = \mathcal{F} \cup \mathcal{G}_1$.

Let $\delta_1 > 0$, $\mathcal{G}_2 \subset C(X)$ and $\mathcal{P}_1 \subset \underline{K}(C(X \times S^1))$ be finite subsets required in 2.5 of [**25**] corresponding to ϵ_1 and $\mathcal{F}_1$. Let $\alpha \in KK(C(X \times S^1), A)$ such that

$$\alpha|_{\ker\rho_{C(X\times S^1)}} = 0 \text{ and } \alpha([1_{C(X)}]) = [1_A].$$

(Note that $\ker\rho_{C(X\times S^1)} = \beta_1(K_1(C(X))).$) Without loss of generality, we may assume that $\mathcal{G}_2 = \mathcal{G} \otimes S$, where $\mathcal{G} \subset C(X)$ is a finite subset. Put $\mathcal{P} = \mathcal{P}_1 \cup \mathcal{P}_0$. We may also assume that $\mathcal{G}_1 \subset \mathcal{G}_2$ and $\delta_1 < \delta_0/2$.

Define $\psi : C(X \times S^1) \to A$ by $\psi(f \otimes g) = h(f)g(u)$ for all $f \in C(X)$ and $g \in C(S^1)$. Then, to simplify the notation, by applying 2.8, we may assume that ψ is a δ_1-$\mathcal{G}_1$-multiplicative contractive completely positive linear map. The condition that $K_1(A) = \{0\}$ and $\mathrm{bott}_1(h, u) = 0$ implies that

$$[\psi]|_{\mathcal{P}} = \alpha|_{\mathcal{P}}.$$

It follows from 2.5 of [**25**] that there exists a unital homomorphism $H_0 : C(X \times S^1) \to A$ such that

$$\|h(f)g(u) - H_0(f \otimes g)\| < \epsilon_1 \text{ for all } f \in \mathcal{F}_1 \tag{e 3.65}$$

and $g \in S$. Let Y be the spectrum of H_0. Then Y is a compact subset of $X \times S^1$. By (3) and (4) of 3.1, the choice of δ_0 and $\mathcal{G}_0$ and the condition that $K_1(A) = \{0\}$, we have $[H_0] = [H_{00}]$ in $KL(C(X \times S^1), A)$ for some point-evaluation $H_{00} : C(X \times S^1) \to A$.

It follows from 1.14 of [**25**] that there is a unital homomorphism $H : C(Y) \to A$ with finite finite dimensional range such that

$$\|H_0(f \otimes g) - H(f \otimes g)\| < \epsilon/4 \text{ for all } f \in \mathcal{F} \tag{e 3.66}$$

and $g = 1$ or $g = z$. In the finite dimensional commutative C^*-subalgebra $H(C(X \times S^1))$, we find a continuous path of unitaries $\{v_t : t \in [0, 1]\}$ such that

$$v_0 = H(1 \otimes z), \ \ v_1 = 1 \text{ and } \mathrm{Length}(\{v_t\}) \le \pi. \tag{e 3.67}$$

The lemma then follows easily. □

Corollary 3.12. *Let $\epsilon > 0$. There is $\delta > 0$ satisfying the following:*

For any two unitaries u and v in a unital C^*-algebra $A \in \mathbf{B}$ with $K_1(A) = \{0\}$ and if

$$\|[u, v]\| < \delta \textit{ and } \mathrm{bott}_1(u, v) = 0,$$

then there exists a continuous path of unitaries $\{u_t : t \in [0, 1]\}$ of A such that

$$u_0 = u, \ \ u_1 = 1 \textit{ and } \|[u_t, v]\| < \epsilon.$$

Moreover,

$$\mathrm{Length}(\{u_t\}) \le \pi + \epsilon.$$

Theorem 3.13. *Let X be a compact metric space with dimension no more than 1. Then, for any $\epsilon > 0$ and any finite subset $\mathcal{F} \subset C(X)$, there exists $\delta > 0$, $\sigma > 0$, a finite subset $\mathcal{G} \subset C(X)$ and a finite subset $\mathcal{P}_0 \subset K_0(C(X))$ and $\mathcal{P}_1 \subset K_1(C(X))$ satisfying the following:*

Suppose that A is a unital simple C^*-algebra in $\mathbf{B}$, suppose that $h : C(X) \to A$ is a unital homomorphism whose spectrum is σ-dense in X and suppose that there is a unitary $u \in A$ such that

$$\|[h(a), u]\| < \delta \textit{ for all } a \in \mathcal{G}, \ \mathrm{bott}_0(h, u)|_{\mathcal{P}_0} = 0 \textit{ and } \mathrm{bott}_1(h, u)|_{\mathcal{P}_1} = 0. \tag{e 3.68}$$

Then there exists a rectifiable continuous path of unitaries $\{u_t : t \in [0,1]\}$ of A such that

$$u_0 = u, \ \ u_1 = 1_A \ \ \text{and} \ \ \|[h(a), u_t]\| < \epsilon \tag{e 3.69}$$

for all $a \in \mathcal{F}$ and all $t \in [0,1]$. Moreover,

$$\text{Length}(\{u_t\}) \le 2\pi + \epsilon. \tag{e 3.70}$$

PROOF. There is (by [**44**]) a sequence of one-dimensional finite CW complexes X_n such that $C(X) = \lim_{n\to\infty}(C(X_n), \phi_n)$. For any $\epsilon > 0$ and any finite subset $\mathcal{F} \subset C(X)$, there is an integer N and a finite subset $\mathcal{F}_1 \subset C(X_N)$ such that for any $f \in \mathcal{F}$, there exists $g_f \in \mathcal{F}_1$ such that

$$\|f - \phi_N(g_f)\| < \epsilon/4 \tag{e 3.71}$$

Replacing h by $h \circ \phi_N$, we see the theorem is reduced to the case that X is a compact subset of a finite CW complex of dimension 1.

Now we assume that X is a compact subset of a finite CW complex of dimension 1. Then, there is a sequence of one-dimensional finite CW complexes $X_n \supset X$ such that $\cap_{n=1}^{\infty} X_n = X$. In particular,

$$\lim_{n\to\infty} \text{dist}(X, X_n) = 0.$$

There exist $\epsilon_1 > 0$ and a finite subset $\mathcal{F}_1 \subset C(X_1)$ such that $\text{bott}_1(h', w)$ is well defined for any unital homomorphism $h' : C(X_1) \to A$ and $w \in U(A)$ for which

$$\|[h'(f), w]\| < \epsilon_1 \ \text{ for all } \ f \in \mathcal{F}_1.$$

Note that, since $X_n \subset X_1$ and X_1 is a finite CW complex of dimension 1, $\text{bott}_1(h'', w)$ is well defined as long as $\|[h''(f), w]\| < \epsilon_1$ for all $f \in s_n(\mathcal{F}_1)$ and for any unital homomorphism $h'' : C(X_n) \to A$, where $s_n : C(X_1) \to C(X_n)$ is the quotient map induced by the embedding from X_n into X_1.

Put $\mathcal{F}_0 = s(\mathcal{F}_1)$, where $s : C(X_1) \to C(X)$ is the quotient map induced by the embedding from X into X_1. For convenience, we may assume that $\epsilon_1 < \epsilon/32$. Let $\mathcal{F}_2 = \mathcal{F} \cup \mathcal{F}_0$ and let $\sigma_1 = \sigma_{X,\mathcal{F}_2,\epsilon_1/4}$. Choose $n \ge 1$ so that $X_n \subset \{x \in X_1 : \text{dist}(x, X) < \sigma_1/4\}$. Note that X_n is a finite CW complex of dimension 1. It then is easy to see that each $f \in \mathcal{F}_2$ can be extended to a function $\tilde{f} \in C(X_n)$ such that

$$|\tilde{f}(t) - f(t')| < \epsilon_1/4 \tag{e 3.72}$$

if $t \in X_n$, $t' \in X$ and $\text{dist}(t, t') < \sigma_1/4$. Put $\widetilde{\mathcal{F}_2} = \{\tilde{f} : f \in \mathcal{F}_2\}$ and $\sigma = \sigma_1/2$. By (e 3.72) and from the discussion in 3.8, we have that

$$\sigma_{X,\mathcal{F}_2,\epsilon_1/4} \le \sigma_{X_n\times S^1,\widetilde{\mathcal{F}_2}\otimes S,\epsilon/32}.$$

Let $s' : C(X_n) \to C(X)$ be the quotient map induced by the embedding from X into X_n. We also note that $s'(\widetilde{\mathcal{F}_2}) = \mathcal{F}_2$.

Now let $\delta > 0$ and $\mathcal{G}_1$ (in place of $\mathcal{G}$) be required by 3.7 for ϵ and $\widetilde{\mathcal{F}_2}$ and X_n. Put $\mathcal{G} = s'(\mathcal{G})$. Choose finite but generating subsets $\mathcal{P}_0' \subset K_0(C(X_n))$ and $\mathcal{P}_1 \subset K_1(C(X_n))$. Define $\mathcal{P}_i = s'_{*i}(\mathcal{P}_i')$, $i = 0, 1$. Suppose that h and u satisfy (e 3.68) for the above δ, $\mathcal{G}$, $\mathcal{P}_0$ and $\mathcal{P}_1$ and assume the spectrum of h is σ-dense in X. Define $\phi = h \circ s'$. Then the spectrum of ϕ is σ_1-dense in X_n. Since $X_n \supset X$ and X_n is a finite CW complex of dimension 1, by (e 3.68), it is clear that we have

$$\|[\phi(f), u]\| < \delta \ \text{ for all } \ f \in \mathcal{G}_1, \ \text{bott}_0(\phi, u) = 0 \ \text{ and } \ \text{bott}_1(\phi, u) = 0.$$

Now 3.7 applies to ϕ (see also 3.8). Note that

$$\|[h(a), u_t]\| < \epsilon$$

for all $a \in \mathcal{F}$, if $\|[\phi(f), u_t]\| < \epsilon$ for $f \in \widetilde{\mathcal{F}}$. The lemma follows. □

Remark 3.14. When the C^*-algebra A is in **B**, both Theorem 3.7 and Theorem 3.13 improve the Basic Homotopy Lemma of [**3**].

(1) The Basic Homotopy Lemma in [**3**] deals with the case that $X = S^1$. Theorem 3.7 allows more general spaces X which is a general one-dimensional finite CW complex and may not be embedded into the $\mathbb{R}^2$. Furthermore, Theorem 3.13 allows any compact metric space with dimension no more than 1. It should be noted that there are spaces such as Hawaii ear ring which are one dimensional but are not compact subsets of any one-dimensional finite CW complex.

(2) The constant δ in Theorem 3.11 only depends on X, ϵ and $\mathcal{F}$. It does not depend on the spectrum of h which is a compact subset of X. In other words, δ is independent of the choices of compact subsets of X. This case is a significant improvement since in the Basic Homotopy Lemma of [**3**] the spectrum of the unitary u is assumed to be δ-dense in S^1 with a possible gap. It should also be noted that if we do not care about the independence of δ of the subsets of X the proofs of this section could be further simplified.

(3) In both Theorem 3.13 and Theorem 3.7, the length of the path $\{u_t : t \in [0,1]\}$ is reduced to $2\pi + \epsilon$. This is less than half of $5\pi + 1$ required by the Basic Homotopy Lemma of [**3**]. The Lemma 3.11 shows that, at least for the case that $K_1(A) = \{0\}$ such as A is a simple AF-algebra, the length of the path can be further reduced to $\pi + \epsilon$ which is the best possible upper bound. However, at this point, we do not known whether the bound $2\pi + \epsilon$ can be further reduced in general.

(4) As indicated in [**3**], the path to the proof of Basic Homotopy Lemma in [**3**] is long. Here we present a shorter route. The idea to establish something like 3.3 and then 3.4 is taken from our earlier result in [**41**] (see also Lemma 2.3 of [**23**]). However, the execution of this idea in this section relies heavily on the results in [**25**]. When the dimension of X becomes more than one, things become much more complicated and some of them may not have been foreseen as we will see in the next few sections.

CHAPTER 2

The Basic Homotopy Lemma for higher dimensional spaces

4. K-theory and traces

Lemma 4.1. *Let C be a unital amenable C^*-algebra and let A be a unital C^*-algebra. Let $\phi : C \otimes C(S^1) \to A$ be a unital homomorphism and let $u = \phi(1 \otimes z)$.*

Suppose that $\psi : C \otimes C(S^1) \to A$ is another homomorphism and $v = \psi(1 \otimes z)$ such that

$$[\phi|_C] = [\psi|_C] \text{ in } KL(C, A), \text{ and } \mathrm{Bott}(\phi|_C, u) = \mathrm{Bott}(\psi|_C, v).$$

Then

$$[\phi] = [\psi] \text{ in } KL(C \otimes C(S^1), A).$$

In particular,

$$\mathrm{Bott}(\phi|_C, u) = 0 \tag{e 4.1}$$

if and only if

$$[\psi_0] = [\phi] \text{ in } KL(C \otimes C(S^1), A) \tag{e 4.2}$$

where $\psi_0(a \otimes f(z)) = \phi(a)f(1_A)$ for all $f \in C(S^1)$.

Moreover, we have the following:

(1) *Suppose that $C = C(X)$ for some connected finite CW complex and suppose that $\mathrm{ker}\rho_{C(X)} = \{0\}$ and $K_1(C(X))$ has no torsion, if $[u] = 0$ in $K_1(A)$ and if*

$$\mathrm{bott}_1(\phi|_{C(X)}, u) = 0, \tag{e 4.3}$$

then

$$[\psi_0] = [\phi] \text{ in } KL(C(X \times S^1), A). \tag{e 4.4}$$

(2) *In the case that $K_0(C)$ has no torsion, or $K_0(A)$ is divisible, and at the same time $K_1(C)$ has no torsion, or $K_1(A)$ is divisible, if*

$$\mathrm{Bott}_i(\phi, u) = 0, \ \ i = 0, 1, \tag{e 4.5}$$

then

$$[\psi_0] = [\phi] \text{ in } KL(C(X \times S^1), A) \tag{e 4.6}$$

(3) *In the case that $K_0(A)$ is divisible and $K_1(A) = \{0\}$ and if*

$$\mathrm{bott}_1(\phi|_{C(X)}, u) = 0 \tag{e 4.7}$$

then

$$[\psi_0] = [\phi] \text{ in } KL(C(X \times S^1), A). \tag{e 4.8}$$

(4) *When $K_1(A) = \{0\}$ and $K_0(C(X))$ has no torsion, and if*

$$\mathrm{bott}_1(\phi|_{C(X)}, u) = 0 \tag{e 4.9}$$

then

$$[\psi_0] = [\phi] \text{ in } KL(C(X \times S^1), A). \tag{e 4.10}$$

PROOF. Let

$$\beta^{(0)} : K_0(C) \to K_1(C \otimes C(S^1)), \beta^{(1)} : K_1(C) \to K_0(C \otimes C(S^1)),$$
$$\beta_k^{(0)} : K_0(C, \mathbb{Z}/k\mathbb{Z}) \to K_1(C \otimes C(S^1), \mathbb{Z}/k\mathbb{Z}) \text{ and}$$
$$\beta_k^{(1)} : K_1(C, \mathbb{Z}/k\mathbb{Z}) \to K_0(C \otimes C(S^1)), \mathbb{Z}/k\mathbb{Z})$$

be defined by the Bott map, $k = 2, 3, \ldots$ (as in 2.10). Put $B = C \otimes C(S^1)$. We have the following two commutative diagrams.

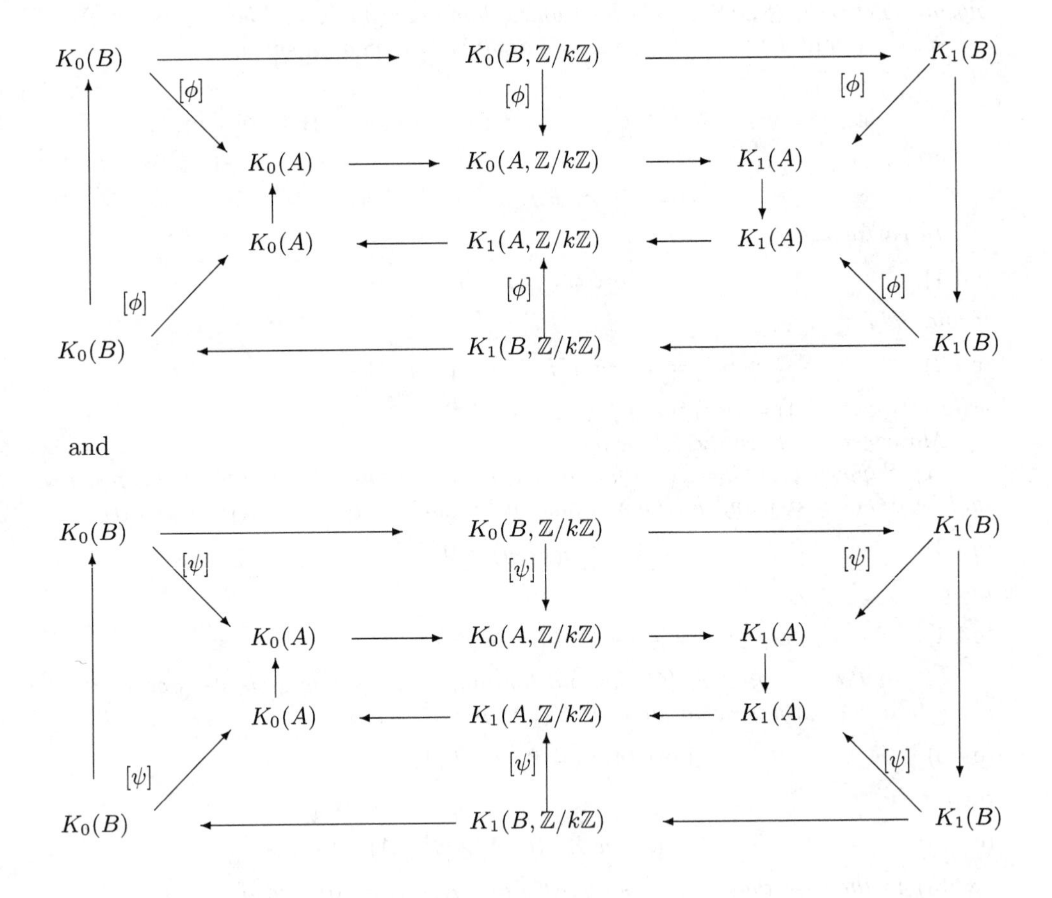

and

$$\begin{array}{ccccc} K_0(B) & \longrightarrow & K_0(B, \mathbb{Z}/k\mathbb{Z}) & \longrightarrow & K_1(B) \\ & \searrow^{[\psi]} & \downarrow [\psi] & {}^{[\psi]}\swarrow & \\ \uparrow & K_0(A) \longrightarrow & K_0(A, \mathbb{Z}/k\mathbb{Z}) & \longrightarrow K_1(A) & \downarrow \\ & \uparrow & & \downarrow & \\ & K_0(A) \longleftarrow & K_1(A, \mathbb{Z}/k\mathbb{Z}) & \longleftarrow K_1(A) & \\ & \nearrow_{[\psi]} & \uparrow [\psi] & {}_{[\psi]}\nwarrow & \\ K_0(B) & \longleftarrow & K_1(B, \mathbb{Z}/k\mathbb{Z}) & \longleftarrow & K_1(B) \end{array}$$

Put $\phi' = \phi|_C$ and $\psi' = \psi|_C$. Then

$$K_i(C \otimes C(S^1) = K_i(C) \bigoplus \beta^{(1-i)}(K_{1-i}(C) \text{ and}$$
$$K_i(C \otimes C(S^1), \mathbb{Z}/k\mathbb{Z}) = K_i(C \otimes C(S^1), \mathbb{Z}/k\mathbb{Z}) \bigoplus \beta_k^{(1-i)}(K_{1-i}(C \otimes C(S^1), \mathbb{Z}/k\mathbb{Z})),$$

$i = 0, 1$ and $k = 2, 3, \ldots$.

This gives the following two commutative diagrams:

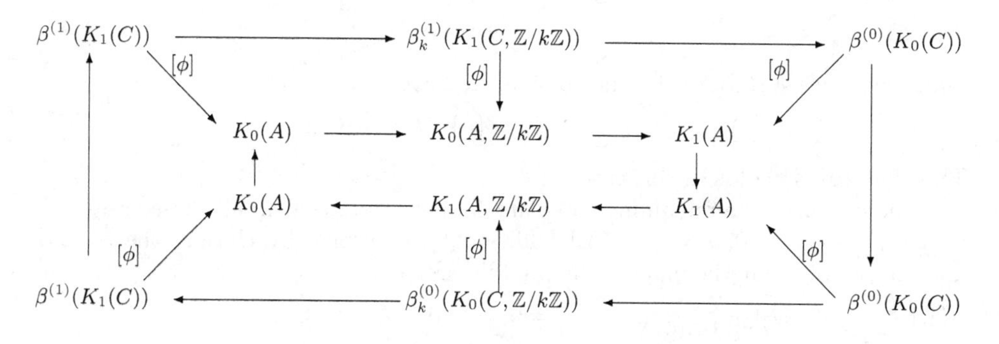

and

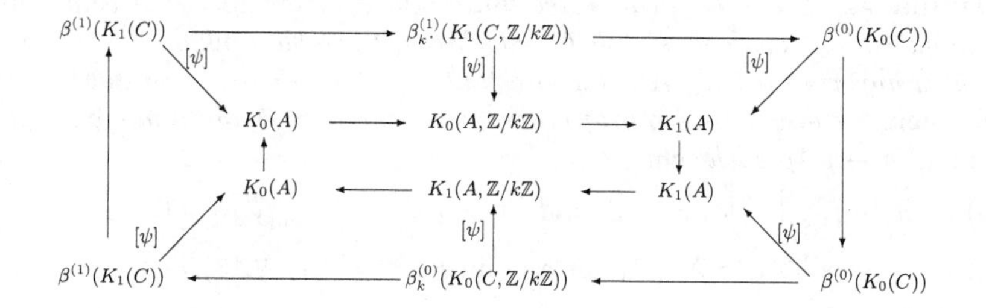

By the assumption that

$$\mathrm{Bott}(\phi', u) = \mathrm{Bott}(\psi', v),$$

all the corresponding maps in the above two diagrams from outside six terms into the inside six terms are the same. Then, by combining the assumption that

$$[\phi'] = [\psi'],$$

we conclude that

$$[\phi] = [\psi].$$

This proves the first part of the statement.

Note that

$$[\psi_0|_C] = [\phi'] \text{ in } KL(C, A) \text{ and}$$
$$\mathrm{Bott}(\psi_0|_C, 1_A) = 0.$$

Thus, by what we have proved,

$$[\psi_0] = [\phi] \text{ in } KL(B, A)$$

if and only if $\mathrm{Bott}(\phi', u) = 0$.

Now consider the special cases.

(1) When $\ker\rho_{C(X)} = \{0\}$, $K_0(C(X)) = \mathbb{Z}$. Thus if $[u] = 0$ in $K_1(A)$,

$$[\phi] \circ \beta^{(0)} = 0 \tag{e 4.11}$$

The condition that $\mathrm{bott}_1(\phi|_{C(X)}, u) = 0$ implies that

$$[\phi] \circ \beta^{(1)} = 0. \tag{e 4.12}$$

Moreover, ker$\rho_{C(X)} = \{0\}$ implies that $K_0(C(X))$ is torsion free. Thus by chasing the upper half of the second diagram above, using the exactness, we see that

$$[\phi]|_{\beta_k^{(1)}(K_1(C(X),\mathbb{Z}/k\mathbb{Z}))} = 0, \;\; k = 2,3,.... \tag{e 4.13}$$

Similarly, since $K_1(C(X))$ has no torsion, we have

$$[\phi]|_{\beta_k^{(0)}(K_0(C(X),\mathbb{Z}/k\mathbb{Z}))} = \{0\}, \;\; k = 2,3,.... \tag{e 4.14}$$

Therefore (e 4.4) holds in this case.

The proofs of the remaining cases are similar. Note that in all these cases, $[\phi]|_{\beta^{(i)}(K_i(C(X))} = 0$, $i = 0,1$. All additional conditions, by chasing the second diagram, imply that the maps in the middle are zero, i.e.,

$$[\phi]|_{\beta_k^{(i)}(K_i(C(X),\mathbb{Z}/k\mathbb{Z}))} = 0, \;\; i = 0,1, k = 2,3,.... \tag{e 4.15}$$

□

Lemma 4.2. *Let X be a connected finite CW complex and let A be an infinite dimensional unital separable simple C^*-algebra with stable rank one, real rank zero and weakly unperforated $K_0(A)$. Suppose that $\phi : C(X) \to A$ is a unital homomorphism. Then, for any non-zero projection $p \in A$, there exist unital monomorphisms $h_1, h_2 : C(X) \to pAp$ such that*

$$[h_1|_{C_0(Y_X)}] = [\phi|_{C_0(Y_X)}] \;\; \textit{and} \;\; [h_2|_{C_0(Y_X)}] + [\phi|_{C_0(Y_X)}] = 0. \tag{e 4.16}$$

Proof. Recall $Y_X = X \setminus \xi_X$ for a point $\xi_X \in X$ (see 2.13).

We have

$$K_0(C(X)) = \mathbb{Z} \bigoplus K_0(C_0(Y_X)) \;\; \text{and} \;\; K_1(C(X)) = K_1(C_0(Y_X)). \tag{e 4.17}$$

Moreover, we may write

$$K_0(C(X), \mathbb{Z}/k\mathbb{Z}) = \mathbb{Z}/k\mathbb{Z} \bigoplus K_0(C_0(Y_X), \mathbb{Z}/k\mathbb{Z}) \;\; \text{and} \tag{e 4.18}$$

$$K_1(C(X), \mathbb{Z}/k\mathbb{Z}) = K_1(C_0(Y_X), \mathbb{Z}/k\mathbb{Z}). \tag{e 4.19}$$

Let $q \le p$ be a nonzero projection such that $p - q \ne 0$. Define an element $\gamma_1 \in Hom_\Lambda(\underline{K}(C(X)), \underline{K}(A))$ by $\gamma_1([1_{C(X)}]) = [q]$ and $\gamma_1(\overline{[1_{C(X)}]}) = \overline{[q]}$, where $\overline{[1_{C(X)}]}$ is the image of $[1_{C(X)}]$ under the map from $K_0(C(X)) \to K_1(C(X), \mathbb{Z}/k\mathbb{Z})$ in $\mathbb{Z}/k\mathbb{Z} \subset K_0(C(X))/kK_0(C(X))$ and $\overline{[q]}$ is the image of $[q]$ under the map $K_0(A) \to K_0(A, \mathbb{Z}/k\mathbb{Z})$ in $K_0(A)/kK_0(A)$, and define $\gamma_1(x) = [\phi](x)$ for $x \in K_0(C_0(Y_X))$, for $x \in K_1(C(X)) = K_1(C_0(Y_X))$, for $x \in K_0(C_0(Y_X), \mathbb{Z}/k\mathbb{Z})$ and for $x \in K_1(C(X), \mathbb{Z}/k\mathbb{Z}) = K_1(C_0(Y_X), \mathbb{Z}/k\mathbb{Z})$, $k = 2,3,....$ It follows from (e 4.18) and (e 4.19) that a straightforward computation shows that $\gamma_1 \in Hom_\Lambda(\underline{K}(C(X)), \underline{K}(A))$. Moreover, γ_1 is an positive element since $[\phi]$ is. Similarly, define $\gamma_2([1_{C(X)}]) = \gamma_1([1_{C(X)}])$, $\gamma_2(\overline{[1_{C(X)}]}) = \gamma_1(\overline{[1_{C(X)}]}) = \overline{[q]}$ and define $\gamma_2(x) = -\gamma_1(x)$ for $x \in K_0(C_0(Y_X))$, for $x \in K_1(C(X)) = K_1(C_0(Y_X))$, for $x \in K_0(C_0(Y_X), \mathbb{Z}/k\mathbb{Z})$ and for $x \in K_1(C(X), \mathbb{Z}/k\mathbb{Z}) = K_1(C_0(Y_X), \mathbb{Z}/k\mathbb{Z})$, $k = 2,3,....$ The same argument shows that $\gamma_2 \in KL(C(X), A)_+$. It now follows from Theorem 4.7 of [**26**] that there are homomorphisms $h_1', h_2' : C(X) \to qAq$ such that

$$[h_1'|_{C_0(Y_X)}] = [\phi|_{C_0(Y_X)}] \;\; \text{and} \;\; [h_2'|_{C_0(Y_X)}] + [\phi|_{C_0(Y_X)}] = 0. \tag{e 4.20}$$

By applying 2.25, we obtain a unital monomorphism $h_{00} : C(X) \to (p-q)A(p-q)$ which factors through $C([0,1])$. In particular,

$$[h_{00}|_{C_0(Y_X)}] = 0 \;\; \text{in} \;\; KK(C(X), A).$$

Now define $h_1 = h_{00} + h_1'$ and $h_2 = h_{00} + h_2''$. □

Lemma 4.3. *Let X be a compact metric space and let A be a unital infinite dimensional separable simple C^*-algebra with real rank zero, stable rank one and weakly unperforated $K_0(A)$. Suppose that $\Lambda : C(X)_{s.a} \to Aff(T(A))$ is a unital positive linear map. Then, for any $\gamma > 0$ and any finite subset $\mathcal{G} \subset C(X)$, there exists a unital homomorphism $\phi : C(X) \to A$ with finite dimensional range such that*

$$|\tau \circ \phi(f) - \Lambda(f)(\tau)| < \gamma \text{ for all } f \in \mathcal{G} \text{ and for all } \tau \in T(A). \tag{e 4.21}$$

The proof is contained in the proof of Theorem 3.6 of [**15**]. We present it here for the convenience of the reader.

PROOF. Let $\gamma > 0$ and $\mathcal{G} \subset C(X)$ be given. Without loss of generality, we may assume that $\mathcal{G} \subset C(X)_{s.a}$. Choose $\delta = \sigma_{X,\mathcal{G},\gamma/3}$. Suppose that $\{y_1, y_2, ..., y_n\}$ is $\delta/2$-dense in X. Let

$$Y_j = \{x \in X : \mathrm{dist}(y, y_j) < \delta\}, j = 1, 2, ..., n.$$

Let $\{g_1, \cdots, g_n\}$ be a partition of unity, i.e., a subset of nonnegative functions in $C(X)$ satisfying the following conditions:
(i) $g_i(x) = 0$ for all $x \notin Y_i, i = 1, 2 \cdots, n$ and
(ii) $\sum_{i=1}^n g_i(x) = 1$ for all $x \in X$.
For each i, define $\hat{g}_i \in \mathrm{Aff}(T(A))$ by

$$\hat{g}_i(\tau) = \tau(\Lambda(g_i)).$$

If $\Lambda(g_i) \neq 0$, then

$$\inf\{\hat{g}_i(\tau) : \tau \in T(A)\} > 0,$$

since A is simple and $T(A)$ is compact. Put

$$\gamma_1 = \inf\{\hat{g}_i(\tau) : \tau \in T(A), g_i \in \mathcal{G}, \Lambda(g_i) \neq 0\}.$$

Let $\gamma_2 = \min\{\gamma, \gamma_1\}$. If $\hat{g}_i = 0$, we choose $q_i = 0$. Since A has real rank zero and the range of the mapping ρ_A is dense in $\mathrm{Aff}(T(A))$, if $\hat{g}_i \neq 0$, there exists a projection $q_i \in A$ such that

$$\|\rho_A(q_i) - (\hat{g}_i - \frac{\gamma_2}{6n})\| < \frac{\gamma_2}{6n}.$$

So for all $\tau \in T(A)$, if $\hat{g}_i \neq 0$,

$$\tau(\Lambda(g_i)) - \frac{\gamma_2}{3n} < \tau(q_i) < \tau(\Lambda(g_i)).$$

Since $\Lambda \neq 0$, this implies for each i,

$$1 - \frac{\gamma_2}{3} = \sum_{i=1}^n \tau(\Lambda(g_i)) - \frac{\gamma_2}{3} < \sum_{i=1}^n \tau(q_i) < \sum_{i=1}^n \tau(\Lambda(g_i)) = 1.$$

Let $Q = \oplus_{i=1}^n q_i$ be a projection in $M_n(A)$. It follows that $\tau(Q) < \tau(1_A)$ for all $\tau \in T(A)$. By the assumption, $M_n(A)$ has the Fundamental Comparison Property. So from $\tau(Q) < \tau(1_A)$ for all $\tau \in T(A)$, one obtains mutually orthogonal projections $q_1', q_2', ..., q_n' \in A$ such that $[q_i'] = [q_i]$, $i = 1, 2, ..., n$. Put $p_i = q_i'$, $i = 1, 2, ..., n-1$ and $p_n = 1 - \sum_{i=1}^{n-1} q_i'$.

For any $\tau \in T(A)$,

$$\tau(\Lambda(g_n)) = 1 - \sum_{i=1}^{n-1} \tau(\Lambda(g_i)) \le 1 - \sum_{i=1}^{n-1} \tau(q_i) = \tau(p_n)$$

and

$$\tau(p_n) = 1 - \sum_{i=1}^{n-1} \tau(q_i) < 1 - (\sum_{i=1}^{n-1} \tau(\Lambda(g_i)) - \frac{\gamma_2}{3}) < \tau(\Lambda(g_n)) + \frac{\gamma_2}{3}.$$

So, for all $\tau \in T(A)$,

$$|\tau(p_i) - \tau(\Lambda(g_i))| \le \frac{\gamma_2}{3n}, \ i = 1, \cdots, n-1,$$

and

$$|\tau(p_n) - \tau(\Lambda(g_n)| < \frac{\gamma_2}{3}.$$

Define homomorphism ϕ by $\phi(f) = \sum_{i=1}^n f(y_i)p_i$ for $f \in C(X)$. Then ψ is a unital homomorphism from $C(X)$ into A with finite dimensional range. Now, we have, for any $\tau \in T(A)$, that

$$\begin{aligned} |\tau(\Lambda(f)) - \tau(\phi(f))| &= |\tau(\Lambda(f)) - \sum_{k=1}^n f(y_i)\tau(p_i)| \\ &\le |\tau(\Lambda(f)) - \sum_{i=1}^n f(y_i)\tau(\Lambda(g_i))| + |\sum_{i=1}^n f(y_i)(\tau(p_i) - \tau(\Lambda(g_i)))| \\ &\le \sum_{i=1}^n \int_{Y_i} |f(x) - f(y_i)| g_i(x) d\mu_{\tau\circ\Lambda} + (n-1)\frac{\gamma_2}{3n} + \frac{\gamma_2}{3} \\ &< \frac{\gamma}{3} + \frac{2\gamma_2}{3} \le \gamma. \end{aligned}$$

□

Corollary 4.4. *Let X be a compact metric space and let A be a unital infinite dimensional separable simple C^*-algebra with real rank zero, stable rank one and weakly unperforated $K_0(A)$. Suppose that $L : C(X) \to A$ is a unital positive linear map. Then, for any $\gamma > 0$ and any finite subset $\mathcal{G} \subset C(X)$, there exists a unital homomorphism $\phi : C(X) \to A$ such that*

(e 4.22) $|\tau \circ \phi(f) - \tau \circ L(f)| < \gamma$ *for all* $f \in \mathcal{G}$ *and for all* $\tau \in T(A)$.

5. Some finite dimensional approximations

Lemma 5.1. *Let X be a compact metric space, let $\epsilon > 0$, $\gamma > 0$ and let $\mathcal{F} \subset C(X)$ be a finite subset. There exists $\delta > 0$ and there exists a finite subset $\mathcal{G} \subset C(X)$ satisfy the following: Suppose that A is a unital C^*-algebra of real rank zero, $\tau : A \to \mathbb{C}$ is a state on A and $\phi : C(X) \to A$ is a unital δ-$\mathcal{G}$-multiplicative contractive completely positive linear map. Then, there is a projection $p \in A$, a unital ϵ-$\mathcal{F}$-multiplicative contractive completely positive linear map $\psi : C(X) \to (1-p)A(1-p)$ and a unital homomorphism $h : C(X) \to pAp$ with finite dimensional range such that*

$$\|\phi(f) - (\psi(f) \oplus h(f))\| < \epsilon \text{ for all } f \in \mathcal{F}$$

and

$$\tau(1-p) < \gamma.$$

PROOF. Fix $\epsilon_0 > 0$ and $\gamma_0 > 0$. Fix a finite subset $\mathcal{F}_0 \subset C(X)$. Suppose that the lemma is false. There exists a sequence of unital C^*-algebras $\{A_n\}$ of real rank zero, a sequence of unital δ_n-$\mathcal{G}_n$-multiplicative contractive completely positive linear maps $\phi_n : C(X) \to A_n$ and a sequence of states τ_n of A_n, where $\sum_{n=1}^{\infty} \delta_n < \infty$ and $\cup_{n=1}^{\infty} \mathcal{G}_n$ is dense in $C(X)$, satisfying the following:

$$\inf\{\sup\{\|\phi_n(f) - [(1-p_n)\phi_n(f)(1-p_n) + h_n(f)]\| : f \in \mathcal{F}_0\}\} \geq \epsilon_0, \tag{e 5.1}$$

where the infimum is taken among all possible projections $p_n \in A_n$ with

$$\tau_n(1-p_n) < \gamma_0 \text{ and}$$

all possible unital homomorphisms $h_n : C(X) \to p_nAp_n$, $n = 1, 2, ...$

Define $\Phi : C(X) \to l^\infty(\{A_n\})$ by $\Phi(f) = \{\phi_n(f)\}$ and define $q : l^\infty(\{B_n\}) \to q_\infty(\{B_n\}) = l^\infty(\{B_n\})/c_0(\{B_n\})$. Then $q \circ \Phi : C(X) \to q^\infty(\{B_n\})$ is a homomorphism. There is a compact subset $Y \subset X$ and monomorphism $\bar{\Psi} : C(Y) \to q_\infty(\{B_n\})$ such that

$$\bar{\Psi} \circ \pi = q \circ \Phi,$$

where $\pi : C(X) \to C(Y)$ is the quotient map. Define

$$t_n(\{a_n\}) = \tau_n(a_n)$$

for $\{a_n\} \in l^\infty(\{B_n\})$. Let τ be a weak limit of $\{\tau_n\}$. By passing to a subsequence, if necessary, we may assume that

$$\lim_{n\to\infty} t_n(\{a_n\}) = \tau(\{a_n\})$$

for all $\{a_n\} \in \Phi(C(X))$. Moreover, since for each $\{a_n\} \in c_0(\{B_n\})$, $\lim_{n\to\infty} t_n(a_n) = 0$, we may view τ as a state of $q_\infty(\{B_n\})$. Note that $q_\infty(\{B_n\})$ has real rank zero. It follows from Lemma 2.11 of [**25**] that there is a projection $\bar{p} \in q_\infty(\{B_n\})$ satisfying the following:

$$\|\bar{\Psi}(f) - [(1-\bar{p})\bar{\Psi}(f)(1-\bar{p}) + \sum_{i=1}^{m} f(x_i)\bar{p}_i]\| < \epsilon_0/3 \text{ and} \tag{e 5.2}$$

$$\|(1-\bar{p})\Psi(f) - \Psi(f)(1-\bar{p})\| < \epsilon_0/3 \tag{e 5.3}$$

for all $f \in \mathcal{F}_0$, where $\{\bar{p}_1, \bar{p}_2, ..., \bar{p}_m\}$ is a set of mutually orthogonal projections and $\{x_1, x_2, ..., x_m\}$ is a finite subset of X, and where

$$\tau(1-\bar{p}) < \gamma_0/3.$$

It follows easily that there is a projection $P \in l^\infty(\{B_n\}))$ and are mutually orthogonal projections $\{P_1, P_2, ..., P_m\} \subset l^\infty(\{B_n\})$ such that

$$q(P) = \bar{p} \text{ and } q(P_i) = \bar{p}_i,\ \ i = 1, 2, ..., m.$$

There are projections $p^{(n)}, p_i^{(n)} \in A_n$, $i = 1, 2, ..., m$, and $n = 1, 2, ...$ such that

$$P = \{p^{(n)}\} \text{ and } P_i = \{p_i^{(n)}\},\ \ i = 1, 2, ..., m.$$

It follows that, for all sufficiently large n,

$$\|\phi_n(f) - [(1-p^{(n)})\phi_n(f)(1-p^{(n)}) - \sum_{i=1}^{m} f(x_i)p_i^{(n)}]\| < \epsilon_0/2$$

for all $f \in \mathcal{F}_0$ and

$$\tau_n(1-p^{(n)}) < \gamma_0/2.$$

This contradicts with (e 5.1). $\square$

Corollary 5.2. *Let X be a compact metric space, let $\epsilon > 0$, $\gamma > 0$ and let $\mathcal{F} \subset C(X)$ be a finite subset. There exists $\delta > 0$ and there exists a finite subset $\mathcal{G} \subset C(X)$ satisfying the following: Suppose that A is a unital C^*-algebra of real rank zero, $\tau : A \to \mathbb{C}$ is a state on A and $\phi : C(X) \to A$ is a unital δ-$\mathcal{G}$-multiplicative contractive completely positive linear map and $u \in A$ is an element such that $\|u^*u - 1\| < \delta$, $\|uu^* - 1\| < \delta$ and*

$$\|[\phi(g), u]\| < \delta \text{ for all } g \in \mathcal{G}.$$

Then, there is a projection $p \in A$, and a unital homomorphism $h : C(X) \to pAp$ with finite dimensional range such that

$$\|\phi(f)g(u) - ((1-p)\phi(f)g(u)(1-p) \oplus h(f \otimes g))\| < \epsilon \text{ and} \tag{e 5.4}$$

$$\|[(1-p), \phi(f)]\| < \delta \tag{e 5.5}$$

for all $f \in \mathcal{F}$ and $g = 1_{C(X)}$ or $g = z$, where

$$h(f \otimes g) = \sum_{i=1}^{n} \sum_{j=1}^{m(i)} f(y_i)g(t_{i,j})p_{i,j} \text{ for all } f \in C(X), g \in C(S^1), \tag{e 5.6}$$

where $y_i \in X$ and $t_{i,j} \in S^1$ and $\{p_{i,j} : i, j\}$ is a set of mutually orthogonal non-zero projections with $\sum_{i=1} \sum_{j=1}^{m(i)} p_{i,j} = p$ and

$$\tau(1-p) < \gamma.$$

Moreover, for any $\eta > 0$, we may assume that

$$\sum_{j=1}^{m(i)} p_{i,j} \subset \overline{\phi(f_i \otimes 1)A\phi(f_i \otimes 1)}, \tag{e 5.7}$$

where $f_i \in C(X_1)$ such that $0 \le f_i \le 1$, $f_i(y) = 1$ if $y \in O(y_i, \eta/2) \subset X_1$ and $f_i(y) = 0$ if $y \notin O(y_i, \eta) \subset X_1$, $i = 1, 2, ..., n$.

Proof. We make a few modification of the proof of 5.1. In the proof of 5.1, define a linear map $L_n : C(X \otimes S^1) \to B_n$ by $L_n(f \otimes g) = \phi_n(f)g(u_n)$, where $\phi_n : C(X) \to B_n$ is a sequence of δ_n-$\mathcal{G}_n$-multiplicative contractive completely positive linear maps and $u_n \in B_n$ with

$$\lim_{n\to\infty} \|u_n^*u_n - 1_{B_n}\| = 0 = \lim_{n\to\infty} \|u_n u_n^* - 1_{B_n}\| = 0 \text{ and}$$
$$\lim_{n\to\infty} \|[\phi_n(f), u_n]\| = 0 \text{ for all } f \in C(X).$$

Redefine $\Phi : C(X \times S^1) \to l^\infty(\{B_n\})$ by $\Psi(f \otimes g) = \{L_n(f \otimes g)\}$ for $f \in C(X)$ and $g \in C(S^1)$. Then $\bar{\Psi} = q\circ\Psi : C(X \times S^1) \to q_\infty(\{B_n\})$ is a unital homomorphism, where $q : l^\infty(\{B_n\}) \to q_\infty(\{B_n\})$ is the quotient map.

In this case, in (e 5.2), we may replace (e 5.2) by the following:

$$\|\bar{\Psi}(f) - [(1-\bar{p})\bar{\Psi}(f)(1-\bar{p}) + \sum_{i=1}^{n} \sum_{j=1}^{m(i)} f(y_i)g(t_{i,j})\bar{p}_{i,j}]\| < \epsilon_0/3 \tag{e 5.8}$$

for all $f \in \mathcal{F}$ and $g \in S$, where $S = \{1_{C(S^1)}, z\}$. As in proof of 5.1, we conclude that (e 5.4) and (e 5.5) hold.

For any $\eta > 0$, from the proof of 2.11 of [**25**], we may assume that

$$\bar{p}_{i,j} \in B_{G_i}, \; j = 1, 2, ..., m(i), i = 1, 2, ..., n \tag{e 5.9}$$

where $G_i = O(y_i, \eta/2)$ and B_{G_i} is the hereditary C^*-subalgebra generated by $\bar{\Psi}(f_i \otimes 1)$, $i = 1, 2, ..., n$.

Let B_n be as in the proof of 5.1. Let $p_{i,j}^{(n)} \in B_n$ be projections such that $q \circ (\{p_{i,j}^{(n)}\}) = \bar{p}_{i,j}$, $j = 1, 2, ..., m(i)$ and $i = 1, 2, ..., n$. Then

(e 5.10) $$\lim_{n\to\infty} \|\phi_n(f_i \otimes 1)p_{i,j}^{(n)}\phi_n(f_i \otimes 1) - p_{i,j}^{(n)}\| = 0.$$

It follows from 2.5.3 of [**29**] that, for each sufficiently large n, there is a projection $q_{i,j}^{(n)} \in \overline{\phi_n(f_i \otimes 1)A\phi_n(f_i \otimes 1)}$ such that

(e 5.11) $$\lim_{n\to\infty} \|p_{i,j}^{(n)} - q_{i,j}^{(n)}\| = 0.$$

It follows that we may replace $p_{i,j}^{(n)}$ by $q_{i,j}^{(n)}$ and then the corollary follows. □

Lemma 5.3. *Let X be a compact metric space, let $\epsilon > 0$, $\gamma > 0$ and let $\mathcal{F} \subset C(X)$ be a finite subset. There exists $\delta > 0$ and there exists a finite subset $\mathcal{G} \subset C(X)$ satisfying the following: Suppose that A is a unital C^*-algebra of tracial rank zero and $\phi : C(X) \to A$ is a unital δ-$\mathcal{G}$-multiplicative contractive completely positive linear map. Then, there is a projection $p \in A$, a unital ϵ-$\mathcal{F}$-multiplicative contractive completely positive linear map $\psi : C(X) \to (1-p)A(1-p)$ and a unital homomorphism $h : C(X) \to pAp$ with finite dimensional range such that*

$$\|\phi(f) - (\psi(f) \oplus h(f))\| < \epsilon \text{ for all } f \in \mathcal{F}$$

and

$$\tau(1-p) < \gamma \text{ for all } \tau \in T(A).$$

In particular, ψ can be chosen to be $\psi(f) = (1-p)\phi(f)(1-p)$ for all $f \in C(X)$.

Proof. Fix $\epsilon > 0$, $\gamma > 0$ and a finite subset $\mathcal{F} \subset C(X)$.

Let $\delta_0 > 0$ and $\mathcal{G}_0 \subset C(X)$ be finite subset required by 5.1 corresponding to $\epsilon/2$, $\gamma/2$ and $\mathcal{F}$.

Choose $\delta = \delta_0/2$ and $\mathcal{G} = \mathcal{G}_0$ and let $\phi : C(X) \to A$ be a homomorphism which satisfies the conditions of the lemma.

Since $TR(A) = 0$, there exists a sequence of finite dimensional C^*-subalgebras B_n with $e_n = 1_{B_n}$ and a sequence of contractive completely positive linear maps $\phi_n : A \to B_n$ such that

(1) $\lim_{n\to\infty} \|e_n a - a e_n\| = 0$ for all $a \in A$,
(2) $\lim_{n\to\infty} \|\phi_n(a) - e_n a e_n\| = 0$ for all $a \in A$ and $\phi_n(1) = e_n$;
(3) $\lim_{n\to\infty} \|\phi_n(ab) - \phi_n(a)\phi_n(b)\| = 0$ for all $a, b \in A$ and
(4) $\tau(1 - e_n) \to 0$ uniformly on $T(A)$.

We write $B_n = \bigoplus_{i=1}^{r(n)} D(i, n)$, where each $D(i, n)$ is a simple finite dimensional C^*-algebra, a full matrix algebra. Denote by $\Phi(i, n) : A \to D(i, n)$ the map which is the composition of the projection map from B_n onto $D(i, n)$ with ϕ_n. Denote by $\tau(i, n)$ the standard normalized trace on $D(i, n)$. Put $\phi_{(i,n)} = \Phi(i, n) \circ \phi$. From (1), (2), (3), (4) above, by applying 5.1 to each $\phi_{(i,n)}$, for each i (and $\tau(i, n)$) and all sufficiently large n, we have

(e 5.12)
$$\|\phi_{(i,n)}(f) - [(1_{D(i,n)} - p_{i,n})\phi_{(i,n)}(f \otimes z)(1_{D(i,n)} - p_{i,n}) + \sum_{j=1}^{m(i)} f(x_{i,j})p_{j,i,n}]\| < \epsilon/4$$

for all $f \in \mathcal{F}$, where $x_j \in X$, $\{p_{j,i,n} : j, i\}$ are mutually orthogonal projections in $D(i,n)$, $\sum_{j=1}^{m(i)} p_{j,i,n} = p_{i,n}$ with

$$\tau(i,n)(e_n - p_{i,n}) < \gamma/3 \tag{e 5.13}$$

and $p_{i,n} = \sum_{j=1}^{m} p_{j,i,n}$, $i = 1, 2, ..., r(n)$.

Put $q_n = \sum_{i=1}^{r(n)} p_{i,n}$. For any $\tau \in T(A)$, $\tau|_{B_n}$ has the form

$$\tau|_{B_n} = \sum_{i=1}^{r(n)} \alpha_{i,n}\tau(i,n),$$

where $\alpha_{i,n} \geq 0$ and $\sum_{i=1}^{r(n)} \alpha_{i,n} < 1$. Thus, by (e 5.13),

$$\tau(e_n - q_n) < \sum_{i=1}^{r(n)} \alpha_{i,n}(\gamma/3) = \gamma/3$$

Therefore

$$\tau(1 - p_n) < \gamma/3 + \gamma/3 < \gamma \text{ for all } \tau \in T(A). \tag{e 5.14}$$

Define $h_n : C(X) \to B_n$ by

$$h_n(f) = \sum_{i=1}^{r(n)} \sum_{j=1}^{m(i)} f(x_{i,j}) p_{j,i,n}$$

for all $f \in C(X)$ and define $\psi_n : C(X) \to (1 - q_n)A(1 - q_n)$ by

$$\psi_n(f) = (1 - q_n)\psi(f)(1 - q_n)$$

for $f \in C(X)$. From (e 5.12) and (e 5.14), the lemma follows by choosing $h = h_n$ and $\psi = \psi_n$ for sufficiently large n and with $\delta < \epsilon/2$ and $\mathcal{G} \supset \mathcal{F}$. □

Lemma 5.4. *Let X be a compact metric space, let $\epsilon > 0$, $\gamma > 0$ and let $\mathcal{F} \subset C(X)$ be a finite subset. There exists $\delta > 0$ and there exists a finite subset $\mathcal{G} \subset C(X)$ satisfying the following: Suppose that A is a unital C^*-algebra of tracial rank zero and $\phi : C(X) \to A$ is a unital homomorphism and $u \in A$ is a unitary such that*

$$\|[\phi(g), u]\| < \delta \text{ for all } g \in \mathcal{G}.$$

Then, for any $\eta > 0$ and any finite subset $\mathcal{F}_1 \subset C(X)$, there is a projection $p \in A$ and a unital homomorphism $h : C(X) \to pAp$ with finite dimensional range such that

$$\|\phi(f)g(u) - ((1-p)\phi(f)g(u)(1-p) \oplus h(f \otimes g))\| < \epsilon \tag{e 5.15}$$

for all $f \in \mathcal{F}, g \in S$,

$$\|\phi(f) - (1-p)\phi(f)(1-p) \oplus h(f \otimes 1)\| < \eta \text{ for all } f \in \mathcal{F}_1, \tag{e 5.16}$$

$$\|[(1-p), \phi(f)]\| < \eta \text{ for all } f \in \mathcal{F}_1 \tag{e 5.17}$$

and

$$\tau(1-p) < \gamma \textit{ for all } \tau \in T(A),$$

where $S = \{1_{C(S^1)}, z\}$.

PROOF. We will use the proof of 5.3 and 5.2. We proceed as in the proof of 5.3. By applying 5.2, instead of (e 5.12) we can have

$$\|\Phi(i,n)(\phi(f)g(u)) \tag{e 5.18}$$

$$-[(1_{D(i,n)}-p_{i,n})\Phi(i,n)(\phi(f)g(u))(1_{D(i,n)}-p_{i,n})+H_{i,n}(f\otimes g)]\|<\epsilon/4 \tag{e 5.19}$$

for all $f\in\mathcal{F}$ and $g\in S$, where

$$H_{i,n}(f\otimes g)=\sum_{j=1}^{L(i)}\sum_{k=1}^{m(j)} f(x_{i,j})g(t_{i,j,k})p_{i,j,k} \tag{e 5.20}$$

for all $f\in C(X)$ and $g\in C(S^1)$, and where $\{p_{i,j,k}: i,j,k\}$ are mutually orthogonal projections with $\sum_{j,k}p_{i,j,k}=p_{i,n}$. For any $\eta>0$, choose $\sigma=\sigma_{X,\mathcal{F}_1,\epsilon/8}$.

By applying 5.2, we may assume that

$$p_{i,j,k}\in\overline{\Phi(i,n)\circ\phi(f_{i,j})A\Phi(i,n)\circ\phi(f_{i,j})}, \tag{e 5.21}$$

where $f_{i,j}\in C(X)$ such that $0\le f_{ij}\le 1$, $f_{i,j}(x)=1$ if $x\in O(x_{i,j},\sigma/2)$ and $f_{i,j}(x)=0$ if $x\notin O(x_{i,j},\sigma)$. By the choice of σ, it follows, since ϕ is a homomorphism, for all sufficiently large n,

$$\|\Phi(i,n)\circ\phi(f)-[(1_{D(i,n)}-p_{i,n})\Phi(i,n)\circ\phi(f)(1_{D(i,n)}-p_{i,n})+H_{i,n}(f\otimes 1)]\|<\eta$$

and

$$\|[\Phi(i,n)\circ\phi(f),\Phi(i,n)(u)]\|<\eta$$

for all $f\in\mathcal{F}_1$. The lemma then follows. □

Corollary 5.5. *Let X be a compact metric space, let $\epsilon>0$ and let $\mathcal{F}\subset C(X)$ be a finite subset. There exists $\delta>0$ and there exists a finite subset $\mathcal{G}\subset C(X)$ satisfying the following: Suppose that A is a unital C^*-algebra of real rank zero and $\phi: C(X)\to A$ is a unital homomorphism and $u\in A$ is a unitary such that*

$$\|[\phi(g),\, u]\|<\delta \text{ for all } g\in\mathcal{G}.$$

Then, for any $\eta>0$ and any finite subset $\mathcal{F}_1\subset C(X)$, there is a non-zero projection $p\in A$ and a unital homomorphism $h: C(X\times S^1)\to pAp$ with finite dimensional range such that

$$\|\phi(f)g(u)-((1-p)\phi(f)g(u)(1-p)\oplus h(f\otimes g))\|<\epsilon \tag{e 5.22}$$

for all $f\in\mathcal{F}, g\in S$,

$$\|\phi(f)-(1-p)\phi(f)(1-p)\oplus h(f\otimes 1)\|<\eta \text{ for all } f\in\mathcal{F}_1, \tag{e 5.23}$$

$$\|[(1-p),\,\phi(f)]\|<\eta \text{ for all } f\in\mathcal{F}_1. \tag{e 5.24}$$

PROOF. This follows from 5.2 as in the proof of 5.4. But we do not need to consider the trace. □

Remark 5.6. It is essentially important in both 5.5 and 5.4 that η and $\mathcal{F}_1$ are independent of the choice of δ and $\mathcal{G}$ so that η can be made arbitrarily small and $\mathcal{F}_1$ can be made arbitrarily large even after δ and $\mathcal{G}$ have been determined.

In 5.5, $h(f)=\sum_{j=1}^m f(\xi_j)p_j$ for all $f\in C(X\times S^1)$, where $\xi_j\in X\times S^1$ and $p_1,p_2,...,p_m$ is a set of mutually orthogonal non-zero projections. In particular, we may assume that $h(f)=f(\xi_1)p_1$, by replacing $(1-p)\phi(1-p)$ by $(1-p)\phi(1-p)+\sum_{j=2}^m f(\xi_j)p_j$. But the latter term is close to $(1-p_1)\phi(1-p_1)$ within 2ϵ in (e 5.22) and within 2η in (e 5.23). Moreover, (e 5.24) holds if one replaces p by p_1 and η by

2η. Therefore, the lemma holds, if h has the form $h(f) = f(\xi)p$ for a single point. Moreover, it works for any projection $q \le p$.

We need the following statement of a result of S. Zhang (1.3 of [**56**]).

Lemma 5.7 (Zhang's Riesz Interpolation). *Let A be a unital C^*-algebra with real rank zero. If $p_1, p_2, ..., p_n$ are mutually orthogonal projections in A and q is another projection such that q is equivalent to a projection in $(p_1+p_2+\cdots p_n)A(p_1+p_2+\cdots p_n)$, then there are projections q' and $q_1, q_2, ..., q_n$ in A such that q and q' are equivalent and $q' = q_1 + q_2 + \cdots + q_n$, $q_i \le p_i$, $i = 1, 2, ..., n$.*

Proof. This follows from 1.3 of [**56**] immediately. There is a partial isometry $v \in A$ such that

$$v^*v = q \text{ and } vv^* \le p_1 + p_2 \cdots + p_n.$$

Put $e = vv^*$. By 1.3 of [**56**], one obtains mutually orthogonal projections $e_1, e_2, ..., e_n$ $\in A$ such that

$$e = e_1 + e_2 \cdots + e_n$$

and there are partial isometries $v_i \in A$ such that

$$v_i^* v_i = e_i \text{ and } v_i v_i^* \le p_i, \quad i = 1, 2, ..., n.$$

Now define $q_i = v_i v_i^*$, $i = 1, 2, ..., n$ and $q' = \sum_{i=1}^n q_i$. □

Lemma 5.8. *Let A be a unital C^*-algebra of real rank zero and let $p_1, p_2, ..., p_n$ be a set of mutually orthogonal projections with $p = \sum_{i=1}^n p_i$. Suppose that $e_1, e_2, ...,$ e_m are mutually orthogonal projections such that $e = \sum_{j=1}^m e_j$ and e is equivalent to p. Then there exists a unital commutative finite dimensional C^*-subalgebra B of pAp which contains $p_1, p_2, ..., p_n$ and $q_1, q_2, ..., q_m$ such that q_j is equivalent to e_j, $j = 1, 2, ..., m$.*

Proof. There is a projection $e_j' \in pAp$ such that e_i' is equivalent to e_j, $i = 1, 2, ..., m$. By Zhang's Riesz interpolation theorem (5.7), there are projections $q_{1,i} \le p_i$, $i = 1, 2, ..., n$ such that q_1 is equivalent to e_1, where $q_1 = \sum_{i=1}^n q_{1,i}$. Note that $q_{1,i}$ commutes with each p_i, $i = 1, 2, ..., n$. We then repeat this argument to e_j', $j = 2, 3, ..., m$ and $p_i - q_{1,i}$, $i = 1, 2, ..., n$. The lemma then follows. □

Lemma 5.9. *Let X be a compact metric space, let $\epsilon > 0$, $\gamma > 0$ and $\mathcal{F} \subset C(X)$ be a finite subset. There exists $\delta > 0$ and there exists a finite subset $\mathcal{G} \subset C(X)$ satisfying the following.*

Suppose that A is a unital separable simple C^*-algebra with tracial rank zero and $\psi_1, \psi_2 : C(X) \to A$ are two δ-$\mathcal{G}$-multiplicative contractive completely positive linear maps. Then

$$\|\psi_i(f) - (\phi_i(f) \oplus h_i(f))\| < \epsilon \text{ for all } f \in \mathcal{F}, \tag{e 5.25}$$

where $\phi_i : C(X) \to (1 - p_i)A(1 - p_i)$ is a unital ϵ-$\mathcal{F}$-multiplicative contractive completely positive linear map and $h_i : C(X) \to p_i A p_i$ is a unital homomorphism with finite dimensional range $(i = 1, 2)$ for two unitarily equivalent projections $p_1, p_2 \in A$ for which

$$\tau(1 - p_i) < \sigma \text{ for all } \tau \in T(A). \tag{e 5.26}$$

PROOF. By applying 5.3 to ψ_1 and ψ_2, we obtain a unital ϵ-$\mathcal{F}$-multiplicative contractive completely positive linear map $\phi_i' : C(X) \to (1-p_i)A(1-p_i)$ and a unital homomorphism with finite dimensional range $h_i' : C(X) \to p_iAp_i$ a unital homomorphism with finite dimensional range ($i = 1, 2$) such that

$$\|\psi_i(f) - (\phi_i'(f) \oplus h_i'(f))\| < \epsilon \text{ for all } f \in \mathcal{F},\ i = 1, 2, \text{ and}$$
$$\tau(1-p_i) < \sigma/2 \text{ for all } \tau \in T(A).$$

Write

$$h_i'(f) = \sum_{k=1}^{m(i)} f(x_k^{(i)})p_k^{(i)} \text{ for all } f \in C(X),$$

where $\{p_k^{(i)} : k = 1, 2, ..., m(i)\}$ is a set of mutually orthogonal projections and $x_k^{(i)} \in X$, $k = 1, 2, ..., m(i)$, $i = 1, 2$. Since A is a simple C^*-algebra with tracial rank zero, there is a projection $q^{(i)} \le p^{(i)}$ such that

$$[q^{(1)}] = [q^{(2)}] \text{ and } \tau(p^{(i)} - q^{(i)}) < \sigma/3$$

for all $\tau \in T(A)$. By Zhang's Riesz interpolation (5.7), there are $e_k^{(i)} \le p_k^{(i)}$ such that

$$[\sum_{k=1}^{m(i)} q^{(i)}e_k^{(i)}] = [q^{(i)}],\ i = 1, 2. \tag{e 5.27}$$

Let $p_i = \sum_{k=1}^{m(i)} e_k^{(i)}$, put

$$\phi_i(f) = \phi_i'(f) + \sum_{k=1}^{m(i)} f(x_k^{(i)})(p_k^{(i)} - e_k^{(i)}) \text{ and}$$
$$h_i(f) = \sum_{k=1}^{m(i)} f(x_k^{(i)})e_k^{(i)}$$

for all $f \in C(X)$, $i = 1, 2$. It is clear that the so defined ϕ_i and h_i satisfy the requirements. □

6. The Basic Homotopy Lemma — full spectrum

Theorem 6.1. *Let X be a connected finite CW complex (with a fixed metric), let A be a unital separable simple C^*-algebra with tracial rank zero and let $h : C(X) \to A$ be a unital monomorphism. Suppose that $u \in A$ is a unitary such that*

$$h(a)u = uh(a) \text{ for all } a \in C(X) \text{ and } \mathrm{Bott}(h, u) = 0. \tag{e 6.1}$$

Suppose also that the homomorphism $\psi : C(X \times S^1) \to A$ induced by $\psi(a \otimes z) = h(a)u$ is injective. Then, for any $\epsilon > 0$ and any finite subset $\mathcal{F} \subset C(X)$, there is a continuous rectifiable path of unitaries $\{u_t : t \in [0, 1]\}$ of A such that

$$\|[h(a), u_t]\| < \epsilon \text{ for all } t \in [0, 1],\ a \in \mathcal{F} \text{ and } u_0 = u,\ u_1 = 1_A. \tag{e 6.2}$$

Moreover,

$$\mathrm{Length}(\{u_t\}) \le \pi + \epsilon\pi. \tag{e 6.3}$$

PROOF. Let $\epsilon > 0$ and $\mathcal{F} \subset C(X)$ be a finite subset. Without loss of generality, we may assume that $1_{C(X)} \in \mathcal{F}$. Let $\mathcal{F}_1 = \{a \otimes b : a \in \mathcal{F}, b = u, b = 1\}$. Let $\eta = \sigma_{X \times S^1, \mathcal{F}_1, \epsilon/16}$ (be as in 2.19). Let $\{x_1, x_2, ..., x_m\}$ be $\eta/2$-dense in $X \times S^1$ and let $s \geq 1$ be an integer such that $O_i \cap O_j = \emptyset$, if $i \neq j$, where

$$O_i = \{x \in X \times S^1 : \operatorname{dist}(x, x_i) < \eta/2s\},\ i = 1, 2, ..., m.$$

Since A is simple, there exists $\sigma > 0$ such that

$$\mu_{\tau \circ \psi}(O_i) \geq 2\sigma\eta, \quad i = 1, 2, ..., m \tag{e 6.4}$$

for all $\tau \in T(A)$. Let $1/2 > \gamma > 0$, $\delta > 0$, $\mathcal{G} \subset C(X \times S^1)$ be a finite subset of $C(X \times S^1)$ and $\mathcal{P} \subset \underline{K}(C(X \times S^1))$ be a finite subset required by Theorem 4.6 of [**33**] associated with $\epsilon/16$, $\sigma/2$ and $\mathcal{F}_1$. We may assume that δ and $\mathcal{G}$ are so chosen that $\overline{\mathcal{Q}_{\delta,\mathcal{G}}} \supset \mathcal{P}$. In other words, for any $\mathcal{G}$-δ-multiplicative contractive completely positive linear map $L : C(X \times S^1) \to B$ (any unital C^*-algebra B), $[L]|_{\mathcal{P}}$ is well-defined.

Choose $x \in X$ and let $\xi = x \times 1$. Let $Y_X = X \setminus \{x\}$. Choose $\eta_1 > 0$ such that $|f(\zeta) - f(\zeta')| < \min\{\delta/2, \epsilon/16\}$ if $\operatorname{dist}(\zeta, \zeta') < \eta_1$ for all $f \in \mathcal{G}$. Let $g \in C(X \times S^1)$ be a nonnegative continuous function with $0 \leq g \leq 1$ such that $g(\zeta) = 1$ if $\operatorname{dist}(\zeta, \xi) < \eta_1/2$ and $g(\zeta) = 0$ if $\operatorname{dist}(\zeta, \xi) \geq \eta_1$. Since A is simple and ψ is injective, $\overline{\psi(g)A\psi(g)} \neq \{0\}$. Since $TR(A) = 0$, A has real rank zero. Choose two non-zero mutually orthogonal and mutually equivalent projections e_1 and e_2 in the simple C^*-algebra $\overline{\psi(g)A\psi(g)}$ $(\neq \{0\})$ for which

$$\tau(e_1 + e_2) < \min\{\gamma/4, \sigma\eta/4\} \text{ for all } \tau \in T(A). \tag{e 6.5}$$

It follows from 4.2 that there are unital homomorphisms $h_1 : C(X) \to e_1Ae_1$ and $h_2 : C(X) \to e_2Ae_2$ such that

$$[h_1|_{C_0(Y_X)}] = [h|_{C_0(Y_X)}] \text{ and } [h_2|_{C_0(Y_X)}] + [h|_{C_0(Y_X)}] = 0. \tag{e 6.6}$$

Define $\psi_1 : C(X \times S^1) \to e_1Ae_1$ by $\psi_1(a \otimes f) = h_1(a)f(1)e_1$ for $a \in C(X)$ and $f \in C(S^1)$ and define $\psi_2 : C(X \times S^1) \to e_2Ae_2$ by $\psi_2(a \otimes f) = h_2(a)f(1)e_2$ for $a \in C(X)$ and $f \in C(S^1)$. Put $e = e_1 + e_2$. Define $\Psi_0 : C(X \times S^1) \to (1-e_1)A(1-e_1)$ by $\Psi_0(a \otimes f) = \psi_2(a \otimes f) \oplus (1-e)h(a)g(u)(1-e)$ for all $a \in C(X)$ and $f \in C(S^1)$ and define $\Psi : C(X) \to A$ by

$$\Psi(b) = \psi_1(a \otimes f) \oplus \Psi_0(b) \text{ for all } b \in C(X \times S^1). \tag{e 6.7}$$

Given the choice of η_1, both Ψ_0 and Ψ are δ-$\mathcal{G}$-multiplicative.

Since $\tau(e_1) < \min\{\gamma/4, \sigma \cdot \eta/4\}$, we also have that

$$\mu_{\tau \circ \Psi}(O_i) \geq 7\sigma\eta/4, \quad i = 1, 2, ..., m. \tag{e 6.8}$$

By 4.4, there is a unital homomorphism $\phi_0 : C(X \times S^1) \to (1-e_1)A(1-e_1)$ with finite dimensional range such that

$$|\tau \circ \phi_0(b) - \tau \circ \Psi_0(b)| < \gamma/2 \text{ for all } b \in \mathcal{G} \text{ and} \tag{e 6.9}$$

$$\mu_{\tau \circ \phi_0}(O_i) \geq \sigma\eta,\ i = 1, 2, ..., m \tag{e 6.10}$$

for all $\tau \in T(A)$.

Note that

$$\psi \approx_{\delta/2} f(\xi)(e_1 + e_2) \oplus (1-e)\psi(1-e) \text{ on } \mathcal{G}. \tag{e 6.11}$$

Thus, by applying 4.1 and by (e 6.6), since $\overline{\mathcal{Q}_{\delta,\mathcal{G}}} \supset \mathcal{P}$,

$$[\Psi_0]|_{\mathcal{P}} = [\phi_0]|_{\mathcal{P}}. \tag{e 6.12}$$

Note that $\tau(1-e_1) \geq 3/4$ and $\frac{\gamma/2}{\tau(1-e_1)} < \gamma$ for all $\tau \in T(A)$. It then follows from Theorem 4.6 of [**33**] and (e 6.9), (e 6.10) and (e 6.12) that there exists a unitary $w_1 \in (1-e_1)A(1-e_1)$ such that

$$\mathrm{ad}\, w_1 \circ \phi_0 \approx_{\epsilon/16} \Psi_0 \ \text{ on } \mathcal{F}_1. \tag{e 6.13}$$

On the other hand, we have

$$\mu_{\tau\circ\Psi}(O_i) \geq 7\sigma\eta/4, \ \text{ for all } \ \tau \in T(A),\ i = 1, 2, ..., m. \tag{e 6.14}$$

Moreover,

$$|\tau \circ \Psi(b) - \tau \circ \psi(b)| < \gamma \ \text{ for all } \ b \in \mathcal{G} \ \text{ and } \ \text{ for all } \ \tau \in T(A). \tag{e 6.15}$$

Furthermore, by 4.1 and by (e 6.6),

$$[\Psi]|_{\mathcal{P}} = [\psi]|_{\mathcal{P}}. \tag{e 6.16}$$

It follows from Theorem 4.6 of [**33**] that there exists a unitary $w_2 \in A$ such that

$$\mathrm{ad}\, w_2 \circ \Psi \approx_{\epsilon/16} \psi \ \text{ on } \mathcal{F}_1 \tag{e 6.17}$$

Let $\omega = \phi_0(1 \otimes z)$. Put $A_0 = \phi_0(C(X \times S^1))$. Since ϕ_0 has finite dimensional range, A_0 is a finite dimensional commutative C^*-algebra. Thus there is a continuous rectifiable path of unitaries $\{\omega_t : t \in [0,1]\}$ in $A_0 \subset (1-e_1)A(1-e_1)$ such that

$$\omega_0 = \omega, \ \omega_1 = 1 - e_1 \ \text{ and } \ \|[\phi_0(a \otimes 1), \omega_t]\| = 0 \tag{e 6.18}$$

for all $a \in C(X)$ and for all $t \in [0,1]$, and

$$\mathrm{Length}(\{\omega_t\}) \leq \pi. \tag{e 6.19}$$

Define

$$U_t = w_2^*(e_1 + w_1^*(\omega_t)w_1)w_2 \ \text{ for all } \ t \in [0,1]. \tag{e 6.20}$$

Clearly $U_1 = 1_A$. Then, by (e 6.13) and (e 6.17),

$$\begin{aligned} \|U_0 - u\| &= \|\mathrm{ad}\, w_2 \circ (\psi_1(1 \otimes z) \oplus \mathrm{ad}\, w_1 \circ \phi_0(1 \otimes z)) - u\| \\ &< \epsilon/16 + \|\mathrm{ad}\, w_2 \circ (\psi_1(1 \otimes z) \oplus \Psi_0(1 \otimes z)) - u\| \\ &< \epsilon/16 + \epsilon/16 + \|\psi(1 \otimes z) - u\| = \epsilon/8. \end{aligned} \tag{e 6.21}$$

We also have, by (e 6.13), (e 6.17) (e 6.21) and (e 6.18),

$$\|[h(a), U_t]\| = \|[\psi(a \otimes 1), U_t]\| < 2(\epsilon/8) + \epsilon/16 = 5\epsilon/16 \tag{e 6.22}$$

for all $a \in \mathcal{F}$. Moreover,

$$\mathrm{Length}(\{U_t\}) \leq \pi. \tag{e 6.23}$$

By (e 6.23) and (e 6.22), we obtain a continuous rectifiable path of unitaries $\{u_t : t \in [0,1]\}$ of A such that

$$u = u_0, \ u_1 = 1_A \ \text{ and } \ \|[h(a), u_t]\| < \epsilon \ \text{ for all } \ t \in [0,1]. \tag{e 6.24}$$

Furthermore,

$$\mathrm{Length}(\{u_t\}) \leq \pi + \epsilon\pi. \tag{e 6.25}$$

□

Lemma 6.2. *Let X be a compact metric space, $\epsilon > 0$ and $\mathcal{F} \subset C(X)$ be a finite subset. Let $L \geq 1$ be an integer and let $0 < \eta \leq \sigma_{X,\mathcal{F},\epsilon/8}$. Then, for any integer $s > 0$, any finite $\eta/2$-dense subset $\{x_1, x_2, ..., x_m\}$ of X for which $O_i \cap O_j = \emptyset$, if $i \neq j$, where*

$$O_i = \{\xi \in X : \mathrm{dist}(\xi, x_i) < \eta/2s\}$$

and any $1/2s > \sigma > 0$, there exist a finite subset $\mathcal{G} \subset C(X \times S^1)$, $\delta > 0$ and an integer $l > 0$ with $8\pi/l < \epsilon$ satisfying the following:

For any unital separable simple C^*-algebra A with tracial rank zero and any δ-$\mathcal{G}$-multiplicative contractive completely positive linear map $\phi : C(X \times S^1) \to A$, if $\mu_{\tau\circ\phi}(O_i \times S^1) > \sigma \cdot \eta$ for all $\tau \in T(A)$ and for all i, then there are mutually orthogonal projections $p_{i,1}, p_{i,2}, ..., p_{m,l}$ in A such that

$$\|\phi(f \otimes g(z)) - ((1-p)\phi(f \otimes g(z))(1-p) + \sum_{i=1}^{m}\sum_{j=1}^{J(i)} f(x_i)g(z_{i,j})p_{i,j})\| < \epsilon$$

$$\text{for all } f \in \mathcal{F} \text{ and } \|(1-p)\phi(f \otimes g) - \phi(f \otimes g)(1-p)\| < \epsilon,$$

where $g = 1_{C(S^1)}$ or $g(z) = z$, $p = \sum_{i=1}^{m} p_i$, $p_i = \sum_{j=1}^{k} p_{i,j}$ and $z_{i,j}$ are points on the unit circle, $1 \leq J(i) \leq l$,

$$\tau(p_k) > \frac{3\sigma}{4} \cdot \eta \quad \text{for all } \tau \in T(A),\ k = 1, 2, ..., m.$$

Proof. There are $f_i \in C(X)$ such that $0 \leq f \leq 1$, $f(x) = 0$ if $x \notin O_i$ and

$$\tau \circ \phi(f_i) > \mu_\phi(O_i) - \sigma \cdot \eta/8 \tag{e 6.26}$$

for all $\tau \in T(A)$, $i = 1, 2, ..., m$. Put $\mathcal{F}_1 = \mathcal{F} \cup \{f_1, f_2, ..., f_m\}$, $S = \{1_{C(S^1)}, z\}$, and $\mathcal{F}' = \mathcal{F}_1 \otimes S$.

Put $\gamma = \sigma\eta/16$. Let $\delta > 0$ and $\mathcal{G} \subset C(X \times S^1)$ be a finite subset which are required by 5.3 for $\min\{\epsilon/2, \gamma/2\}$ (in place of ϵ), γ and $\mathcal{F}'$ (in place of $\mathcal{F}$).

Suppose that ϕ satisfies the assumption of the lemma for δ and $\mathcal{G}$ above.

Applying 5.3, we obtain a projection p, a homomorphism $H : C(X \times S^1) \to pAp$ with finite dimensional range and an $\epsilon/2$-$\mathcal{G}$-multiplicative contractive completely positive linear map $\psi_0 : C(X \times S^1) \to (1-p)A(1-p)$ such that

$$\|\phi(f) - [\psi_0(f) + H(f)]\| < \min\{\epsilon/2, \gamma/2\} \tag{e 6.27}$$

for all $f \in \mathcal{F}'$ and

$$\tau(1-p) < \gamma \text{ for all } \tau \in T(A). \tag{e 6.28}$$

Clearly we may assume that $\psi_0(f) = (1-p)\phi(f)(1-p)$ for all $f \in C(X)$. Define $h(f) = H(f \otimes 1)$ for $f \in C(X)$. Then we may write that

$$h(f) = \sum_{j=1}^{k} f(y_j)e_j \text{ for all } f \in C(X),$$

where $y_j \in X$ and $\{e_1, e_2, ..., e_k\}$ is a set of mutually orthogonal projections. In particular,

$$h(f_i) = \sum_{y_j \in O_j} f(y_j)e_j, \quad i = 1, 2, ..., m. \tag{e 6.29}$$

Put $p_i' = \sum_{y_j \in O_j} e_j$. It follows from (e 6.26), (e 6.27), (e 6.28) and (e 6.29) that

$$\tau(p_i') > \frac{3}{4}\sigma \cdot \eta \text{ for all } \tau \in T(A), \ i = 1, 2, ..., m. \tag{e 6.30}$$

Note that, if $d(x, x') < \eta$,

$$|f(x) - f(x')| < \epsilon/8$$

for all $f \in \mathcal{F}$. It is then easy to see (by the choice of l) that we may assume that

$$H(f \otimes g) = \sum_{i=1}^{m} \sum_{j=1}^{J(i)} f(x_i) g(z_{i,j}) p_{i,j}$$

for all $f \in C(X)$ and $g \in C(S^1)$ as required (with $J(i) \leq l$). □

Remark 6.3. In Theorem 6.2, the integer l depends only on ϵ and $\mathcal{F}$. In fact, it is easy to see that l can be chosen so that $\pi M/l < \epsilon/2$, where $M = \max\{\|f\| : f \in \mathcal{F}\}$.

Lemma 6.4. *Let X be a compact metric space, $\epsilon > 0$ and $\mathcal{F} \subset C(X)$ be a finite subset. Let l be a positive integer for which $256\pi M/l < \epsilon$, where $M = \max\{1, \max\{\|f\| : f \in \mathcal{F}\}\}$. Let $\eta = \sigma_{X,\mathcal{F},\epsilon/32}$ be as 2.19. Then, for any finite $\eta/2$-dense subset $\{x_1, x_2, ..., x_m\}$ of X for which $O_i \cap O_j = \emptyset$, where*

$$O_i = \{x \in X : \text{dist}(x, x_i) < \eta/2s\}$$

for some integer $s \geq 1$ and for any $\sigma_1 > 0$ for which $\sigma_1 < 1/2s$, and for any $\delta_0 > 0$ and any finite subset $\mathcal{G}_0 \subset C(X \otimes S^1)$, there exist a finite subset $\mathcal{G} \subset C(X)$, $\delta > 0$ satisfying the following:

Suppose that A is a unital separable simple C^*-algebra with tracial rank zero, $h : C(X) \to A$ is a unital monomorphism and $u \in A$ is a unitary such that

$$\|[h(a), u]\| < \delta \text{ for all } a \in \mathcal{G} \text{ and } \mu_{\tau \circ h}(O_i) \geq \sigma_1 \eta \text{ for all } \tau \in T(A). \tag{e 6.31}$$

Then there is a δ_0-$\mathcal{G}_0$-multiplicative contractive completely positive linear map $\phi : C(X) \otimes C(S^1) \to A$ and a rectifiable continuous path $\{u_t : t \in [0, 1]\}$ such that

$$u_0 = u, \ \|[\phi(a \otimes 1), u_t]\| < \epsilon \text{ for all } a \in \mathcal{F}, \tag{e 6.32}$$

$$\|\phi(a \otimes 1) - h(a)\| < \epsilon, \ \|\phi(a \otimes z) - h(a)u\| < \epsilon \text{ for all } a \in \mathcal{F}, \tag{e 6.33}$$

where $z \in C(S^1)$ is the standard unitary generator of $C(S^1)$, and

$$\mu_{\tau \circ \phi}(O(x_i \times t_j)) > \frac{\sigma_1}{2l}\eta, \ i = 1, 2, ..., m, j = 1, 2, ..., l \tag{e 6.34}$$

for all $\tau \in T(A)$, where $t_1, t_2, ..., t_l$ are l points on the unit circle which divide S^1 into l arcs evenly and where

$$O(x_i \times t_j) = \{x \times t \in X \times S^1 : \text{dist}(x, x_i) < \eta/2s \text{ and } \text{dist}(t, t_j) < \pi/4sl\}$$

for all $\tau \in T(A)$ (so that $O(x_i \times t_j) \cap O(x_{i'} \times t_{j'}) = \emptyset$ if $(i, j) \neq (i', j')$).

Moreover,

$$\text{Length}(\{u_t\}) \leq \pi + \epsilon\pi. \tag{e 6.35}$$

PROOF. Fix ϵ and $\mathcal{F}$. Without loss of generality, we may assume that $\mathcal{F}$ is in the unit ball of $C(X)$. We may assume that $\epsilon < 1/5$ and $1_{C(X)} \in \mathcal{F}$. Let $\eta > 0$ such that $|f(x) - f(x')| < \epsilon/32$ for all $f \in \mathcal{F}$ if $\text{dist}(x, x') < \eta$. Let $\{x_1, x_2, ..., x_m\}$ be an $\eta/2$-dense set. Let $s > 0$ such that $O_i \cap O_j = \emptyset$ if $i \neq j$, where

$$O_i = O(x_i, \eta/2s) = \{x \in X : \text{dist}(x_i, x) < \eta/2s\}, \ i = 1, 2, ..., m. \tag{e 6.36}$$

Suppose also $1/2s > \sigma_1 > 0$ and $\mu_{\tau\circ\phi}(O_i) > \sigma_1\eta$ for all $\tau \in T(A)$. Let δ_0 and $\mathcal{G}_0$ be given. Without loss of generality (by using smaller δ_0), we may assume that $\mathcal{G}_0 = \mathcal{F}_0 \otimes S$, where $\mathcal{G}_0 \subset C(X)$ is a finite subset and $S = \{1_{C(S^1)}, z\}$. Put $\epsilon_1 = \min\{\epsilon, \delta_0\}$ and $\mathcal{F}_1 = \mathcal{F}_0 \cup \mathcal{F}$. We may assume that $\mathcal{F}_1$ is a subset of the unit ball.

Let $\delta_1 > 0$ and $\mathcal{G}_1 \subset C(X)$ be a finite subset (and $8\pi/l < \epsilon/32$) as required by 6.2 corresponding to $\epsilon_1/32$, $\mathcal{F}_1$, s and σ_1 above (instead of ϵ, $\mathcal{F}$, s and σ). We may assume that $\delta_1 < \epsilon/32$ and $1_{C(X)} \in \mathcal{G}_1$. We may further assume that

$$\delta_1/2 \le \delta_0 \text{ and } \mathcal{F}_0 \subset \mathcal{G}_1. \tag{e 6.37}$$

Let $B = C(X)$. Let $\mathcal{G}_2$ a finite subset (in place of $\mathcal{F}_1$) and $\delta_2 > 0$ (in place of δ) be required by Lemma 2.8 associated with $\delta_1/2$ and $\mathcal{G}_1$ (in place of ϵ and $\mathcal{F}_0$).

Suppose that A is a unital separable simple C^*-algebra with tracial rank zero, suppose that $h : C(X) \to A$ is a unital monomorphism, suppose that $u \in A$ is a unitary such that

$$\|[h(a), u]\| < \delta_2 \text{ for all } a \in \mathcal{G}_2 \text{ and } \mu_{\tau\circ h}(O_i) > \sigma_1\eta \tag{e 6.38}$$

for all $\tau \in T(A)$, $i = 1, 2, ..., m$.

We may assume that $\delta_2 < \delta_1$ and $\mathcal{G}_1 \subset \mathcal{G}_2$. It follows from 2.8 that there exists a δ_1-$\mathcal{G}_1$-multiplicative contractive completely positive linear map $\psi : C(X) \otimes C(S^1) \to A$ such that

$$\|\psi(a) - h(a)\| < \delta_1/2 \text{ and } \|\psi(a \otimes z) - h(a)u\| < \delta_1/2 \tag{e 6.39}$$

for all $a \in \mathcal{G}_1$.

By applying 6.2, we obtain a projection $p \in A$ and mutually orthogonal projections $\{p_{i,j} : j = 1, 2, ..., J(i), i = 1, 2, ..., m\}$ with $\sum_{i,j} p_{i,j} = p$ such that $J(i) \le l$,

$$\|\psi(f \otimes g) - [(1-p)\psi(f \otimes g)(1-p) + \sum_{i=1}^{m}\sum_{j=1}^{J(i)} f(x_i)g(z_{i,j})p_{i,j}]\| < \epsilon_1/32 \tag{e 6.40}$$

for all $f \in \mathcal{F}_1$ and for $g \in S$ and

$$\tau(p_i) > 3\sigma_1 \cdot \eta/4 \text{ for all } \tau \in T(A), \tag{e 6.41}$$

where $p_i = \sum_{j=1}^{J(i)} p_{i,j}$, $i = 1, 2, ..., m$.

Since $\epsilon < 1/5$, by (e 6.40), (see for example Lemma 2.5.8 of [**29**]) there are unitaries $w_1 \in (1-p)A(1-p)$, such that

$$\|\psi(1 \otimes z) - w_1 \oplus \sum_{i=1}^{m}\sum_{j=1}^{J(i)} z_{i,j}p_{i,j}\| < \epsilon/32 + \epsilon/4 \tag{e 6.42}$$

for all $f \in \mathcal{F}_1$. Fix i. For each j, since p_iAp_i is a unital simple C^*-algebra with real rank zero and stable rank one and $K_0(A)$ is weakly unperforated, it follows from 5.8 that there is a finite dimensional commutative unital C^*-subalgebra $B_i \subset p_iAp_i$ which contains projections $p_{i,j}$, $j = 1, 2, ..., J(i)$, and mutually orthogonal projections $q_{i,j} \in B_i$, $j = 1, 2, ..., l$, such that $\sum_{j=1}^{l} q_{i,j} = p_i$ and

$$\tau(q_{i,j}) > \frac{\tau(p_i)}{l} - \frac{\sigma_1\eta}{16 \cdot l^2} \text{ for all } \tau \in T(A), \tag{e 6.43}$$

$i = 1, 2, ..., m$. Define $B = \bigoplus_{i=1}^m B_i$ and define a unital homomorphism $\Psi_0 : C(X \times S^1) \to B$ such that

$$\Psi_0(f \otimes g) = \sum_{i=1}^m f(x_i) \sum_{j=1}^l g(t_j) q_{i,j} \text{ for all } f \in C(X) \text{ and } g \in C(S^1). \tag{e 6.44}$$

There is a continuous path of unitaries $\{v_t : t \in [0,1]\}$ in B such that

$$v_0 = w_1 \oplus \sum_{i=1}^m \sum_{j=1}^l t_{i,j} q_{i,j}, \quad v_1 = \Psi_0(1 \otimes z) \text{ and } \mathrm{Length}(\{v_t\}) \le \pi. \tag{e 6.45}$$

Define $U_t = w_1 \oplus v_t$ for $t \in [0,1]$ and define

$$\phi(f \otimes g(z)) = (1-p)\psi(f \otimes z)(1-p) + \sum_{i=1}^m \sum_{j=1}^l f(x_i) g(t_j) q_{i,j} \tag{e 6.46}$$

for all $f \in C(X)$ and $g \in C(S^1)$. By (e 6.40), we know that ϕ is δ_0-$\mathcal{G}_1$-multiplicative.

We compute that

$$U_0 = w_1 \oplus v_0, U_1 = w_1 \oplus v_1, \tag{e 6.47}$$

$$\|U_0 - u\| \le \|U_0 - \psi(1 \otimes z)\| + \|\psi(1 \otimes z) - u\| \tag{e 6.48}$$

$$< \epsilon/32 + \epsilon/4 + \delta_1 < \epsilon/16 + \epsilon/4 \tag{e 6.49}$$

(by (e 6.39) and e 6.42). Also, by (e 6.39) and (e 6.40),

$$\|\phi(a \otimes 1) - h(a)\| < \epsilon/2 \text{ and } \|\phi(a \otimes z) - h(a)U_1\| < \epsilon/2 \tag{e 6.50}$$

for all $a \in \mathcal{F}$, and by (e 6.42),

$$\|[U_t, \phi(f \otimes 1)]\| < \epsilon/32 + \epsilon/4 + \epsilon/32 \text{ for all } t \in [0,1]. \tag{e 6.51}$$

Moreover, by (e 6.53),

$$\mathrm{Length}(\{U_t\}) \le \pi. \tag{e 6.52}$$

Furthermore, by (e 6.43) and (e 6.41) we have

$$\tau(q_{i,s}) > 3\sigma_1 \cdot \eta/4l - \sigma_1 \cdot \eta/16l \ge \sigma_1 \cdot \eta/2l \tag{e 6.53}$$

for all $\tau \in T(A)$ and

$$\mu_{\tau \circ \phi}(O(x_i \times t_j)) \ge \frac{\sigma_1}{2l}\eta, \tag{e 6.54}$$

$i = 1, 2, ..., m$ and $j = 1, 2, ..., l$. By (e 6.48), we may write

$$u(U_0)^* = \exp(ib) \text{ for some } b \in A_{s.a} \text{ with } \|b\| \le \epsilon\pi. \tag{e 6.55}$$

We define a continuous rectifiable path of unitaries $\{u_t : t \in [0,1]\} \subset A$ by

$$u_t = exp(i(1-2t)b)U_0 \text{ for } t \in [0, 1/2] \text{ and } u_t = U_{2t-1}. \tag{e 6.56}$$

So $u_0 = u$ and $u_1 = U_1$. Moreover, by (e 6.51), (e 6.48) and (e 6.55),

$$\|[\phi(f \otimes 1), u_t]\| < \epsilon \text{ for all } t \in [0,1]. \tag{e 6.57}$$

By (e 6.55),

$$\mathrm{Length}(\{u_t : t \in [0, 1/2]\}) \le \epsilon\pi. \tag{e 6.58}$$

Finally, by (e 6.52) and (e 6.58), we obtain that

$$\mathrm{Length}(\{u_t\}) \le \pi + \epsilon\pi. \tag{e 6.59}$$

□

7. The Basic Homotopy Lemma — finite CW complexes

Definition 7.1. Let X be a compact metric space, let $\varDelta : (0,1) \to (0,1)$ be an increasing map and let μ be a Borel probability measure. We say μ is $\varDelta$-distributed, if for any $\eta \in (0,1)$,

$$\mu(O(x,\eta)) \ge \varDelta(\eta)\eta \text{ for all } x \in X.$$

Recall that a Borel measure on X is said to be *strictly positive* if for any non-empty open subset $O \subset X$, $\mu(O) > 0$. If μ is a strictly positive Borel probability measure, then there is always an increasing map $\varDelta : (0,1) \to (0,1)$ such that μ is $\varDelta$-distributed. To see this, fix $\eta \in (0,1)$. Note that there are $x_1, x_2, ..., x_m \in X$ such that

$$\cup_{i=1}^m O(x_i, \eta/2) \supset X.$$

Let

$$\varDelta'(\eta) = \frac{\min\{\mu(O(x_i,\eta/2)) : i = 1, 2, ..., m\}}{\eta}. \tag{e 7.1}$$

Then, for any $x \in X$, there is i such that $x \in O(x_i, \eta/2)$. Thus

$$O(x,\eta) \supset O(x_i, \eta/2). \tag{e 7.2}$$

Therefore

$$\mu(O(x,\eta)) \ge \varDelta'(\eta)\eta \text{ for all } x \in X. \tag{e 7.3}$$

Then, for each $1 > \eta' \ge \eta > 0$ and $x \in X$,

$$\mu(O(x,\eta'/2)) \ge \mu(O(x,\eta/2)) \ge \varDelta'(\eta/2)\eta/2. \tag{e 7.4}$$

It follows from (e 7.4) and (e 7.1) that, for any $1 > \eta' \ge \eta > 0$,

$$\eta'\varDelta'(\eta') \ge \varDelta'(\eta/2)\eta/2.$$

So $\varDelta'(\eta') \ge \varDelta'(\eta/2)\eta/2$ (since $1 > \eta' \ge \eta > 0$). Define

$$\varDelta(\eta) = \min\{\inf\{\varDelta'(\eta') : \eta' \ge \eta\}, 3/4\}. \tag{e 7.5}$$

Then $\varDelta$ is increasing function on $(0,1)$. Moreover, for any $1 > \eta > 0$,

$$\varDelta'(\eta) \ge \varDelta(\eta) = \min\{\inf\{\varDelta(\eta') : \eta' \ge \eta\}.3/4\} \ge \min\{\varDelta'(\eta/2)\eta/2, 3/4\} > 0.$$

So $\varDelta : (0,1) \to (0,1)$ is an increasing function and μ is $\varDelta$-distributed.

Proposition 7.2. *Let A be a unital simple C^*-algebra with tracial state space $T(A)$. Let $\phi : C(X) \to A$ be a monomorphism. Then there is an increasing map $\varDelta : (0,1) \to (0,1)$ such that $\mu_{\tau\circ\phi}$ is $\varDelta$-distributed for all $\tau \in T(A)$.*

Proof. Fix $\eta \in (0,1)$. For each $x \in X$, define $f_{x,\eta} \in C(X)_+$ with $0 \le f_{x,\eta} \le 1$ such that $f_{x,\eta}(x) = 1$ if $x \in O(x,\eta/4)$ and $f_{x,\eta}(x) = 0$ if $x \notin O(x,\eta/2)$. Since A is simple, there exists $d(x,\eta) > 0$ such that

$$\tau(\phi(f_{x,\eta})) > d(x,\eta) \text{ for all } \tau \in T(A).$$

By the compactness of X, there are $x_1, x_2, ..., x_m \in X$ such that

$$\cup_{i=1}^m O(x_i, \eta/2) \supset X.$$

Define

$$D(\eta) = \frac{\min\{d(x_i, \eta/2) : i = 1, 2, ..., m\}}{\eta}$$

for $1 > \eta > 0$. For any $x \in X$, there exists i such that $x \in O(x_i, \eta/2)$. Thus $O(x,\eta) \supset O(x_i, \eta/2)$ for some $1 \le i \le m$. Thus

$$\mu_{\tau\circ\phi}(O(x,\eta)) > D(\eta)\eta$$

for all $x \in X$ and all $\tau \in T(A)$. Define

$$\varDelta = \min\{\inf\{D(\eta') : \eta' \ge \eta\}, 3/4\}.$$

It follows from the discussion right after 7.1 that there is an increasing map $\varDelta : (0,1) \to (0,1)$ such that $\mu_{\tau\circ\phi}$ is $\varDelta$-distributed for all $\tau \in T(A)$. □

Definition 7.3. Let $\epsilon > 0$ and let $\mathcal{F} \subset C(X)$ be a finite subset. Let $1 > \Delta > 0$ be a positive number. We say that a Borel probability measure is $(\epsilon, \mathcal{F}, \mathcal{D}, \Delta)$-distributed, if there exists an $\eta/2$-dense finite subset $\mathcal{D} = \{x_1, x_2, ..., x_m\} \subset X$, where $0 < \eta = \eta_d(\epsilon, \mathcal{F}) = \min\{\frac{\epsilon}{2+\max\{\|f\|: f\in\mathcal{F}\}}, \sigma_{X,\mathcal{F},\epsilon/32}\}$ and an integer $s \ge 1$ satisfies the following:

(1) $O_i \cap O_j = \emptyset$, if $i \ne j$,

(2) $\mu(O_i) \ge \eta \cdot \varDelta$, $i = 1, 2, ..., m$, where

$$O_i = \{x \in X : \mathrm{dist}(x, x_i) < \eta/2s\},\ i = 1, 2, ..., m.$$

Here we assume that

$$\eta/s \le \min\{\mathrm{dist}(x_i, x_j), i \ne j\}.$$

If there is an increasing map $\varDelta : (0,1) \to (0,1)$ such that μ is $\varDelta$-distributed, then, for any $\epsilon > 0$ and any finite subset $\mathcal{F} \subset C(X)$, there exists $\eta > 0$, a finite subset $\mathcal{D}$ which is $\eta/2$ -dense, an integer $s \ge 1$ such that μ is $(\epsilon, \mathcal{F}, \mathcal{D}, \varDelta(\eta/2s)/2s)$-distributed.

To see this, let $\mathcal{D} = \{x_1, x_2, ..., x_m\} \subset X$ be an $\eta/2$-dense subset, where $\eta = \eta_d(\epsilon, \mathcal{F})$ as above and $s \ge 1$ satisfying (1) above. Then

$$\mu(O_i) \ge \Delta(\eta/2s)\eta/2s = (\Delta(\eta/2s)/2s)\eta,\ \ i = 1, 2, ..., m. \tag{e 7.6}$$

Thus μ is $(\epsilon, \mathcal{F}, \mathcal{D}, \varDelta(\eta/2s)/2s)$-distributed.

If $\mathcal{D} = \{x_1, x_2, ..., x_m\}$ is $\eta/2$-dense as above and

$$\min\{\mathrm{dist}(x_i, x_j) : i \ne j\} > \eta/s,$$

then we say that $\mathcal{D}$ is (s, η)-separated.

Theorem 7.4. *Let X be a finite CW-complex with a fixed metric. For any $\epsilon > 0$, any finite subset $\mathcal{F} \subset C(X)$, $0 < \varDelta < 1$ and $0 < \eta < \eta_d(\epsilon, \mathcal{F})$ (see 7.1), there exists $\delta > 0$ and a finite subset $\mathcal{G} \subset C(X)$ satisfying the following:*

Suppose that A is a unital separable simple C^*-algebra with tracial rank zero, $h : C(X) \to A$ is a unital monomorphism and $u \in A$ is a unitary such that $\mu_{\tau\circ h}$ is $(\epsilon, \mathcal{F}, \mathcal{D}, \varDelta)$-distributed for some finite ($\eta/2$-dense) subset $\mathcal{D}$ which is (s, η)-separated (for some integer $s \ge 1$) and for all $\tau \in T(A)$,

$$\|[h(a), u]\| < \delta \text{ for all } a \in \mathcal{G} \text{ and } \mathrm{Bott}(h, u) = 0. \tag{e 7.7}$$

Then, there exists a continuous rectifiable path of unitaries $\{u_t : t \in [0,1]\}$ of A such that

$$u_0 = u,\ \ u_1 = 1_A \text{ and } \|[h(a), u_t]\| < \epsilon \text{ for all } a \in \mathcal{F} \text{ and } t \in [0,1]. \tag{e 7.8}$$

Moreover,

$$\mathrm{Length}(\{u_t\}) \le 2\pi + \epsilon\pi. \tag{e 7.9}$$

PROOF. We first show that we may reduce the general case to the case that X is a connected finite CW complex. We may assume that $X = \cup_{i=1}^k X_i$ which is a disjoint union of connected finite CW complex X_i. Let e_i be the function in $C(X)$ which is 1 in X_i and zero elsewhere. Suppose that we have shown the theorem holds for any connected finite CW complex.

Let δ_i be required for X_i, $i = 1, 2, ..., k$. Put $\delta = \min\{\delta_i/2 : i = 1, 2, ..., k\}$. Put $P_i = h(e_i)$. Then with a sufficiently large $\mathcal{G}$, we may assume that $\|[P_i, u]\| < \delta$. There is, for each i, a unitary $u_i \in P_iAP_i$ such that

$$\|P_iuP_i - u_i\| < 2\delta \text{ and } \|u - \sum_{i=1}^n u_i\| < 2\delta.$$

Note that the condition $\mathrm{Bott}(h, u) = 0$ implies that

$$\mathrm{Bott}(h_i, u_i) = 0, \quad i = 1, 2, ..., k,$$

provided that δ is small enough. These lines of argument lead to the reduction of the general case to the case that X is a connected finite CW complex.

For the rest of the proof, we will assume that X is a connected finite CW complex.

Let $\epsilon > 0$ and $\mathcal{F} \subset C(X)$ be a finite subset. We may assume that $1 \in \mathcal{F}$. Let

$$\mathcal{F}' = \{f \otimes a : f \in \mathcal{F} \text{ and } a = z, \text{ or } a = 1\} \subset C(X \times S^1).$$

Note that

$$0 < \eta < \min\{\frac{\epsilon}{1 + \max\{\|f\| : f \in \mathcal{F}\}}, \sigma_{X,\mathcal{F},\epsilon/32}\}.$$

It follows that if $g \in \mathcal{F}'$, then

$$|g(z) - g(z')| < \epsilon/32 \text{ for all } g \in \mathcal{F}'$$

if $\mathrm{dist}(z, z') < \eta$ $(z, z' \in X \times S^1)$. Let $s \geq 1$ be given and $\Delta > 0$.

To simplify notation, without loss of generality, we may assume that $\mathcal{F}$ is in the unit ball of $C(X)$. Choose l to be an integer such that $256\pi/l < \min\{\eta/4s, \epsilon/16\}$. Let

$$\sigma = \Delta/8l.$$

Let $\gamma > 0$, $\delta_1 > 0$, $\mathcal{G}_1 \subset C(X \times S^1)$ be a finite subset and $\mathcal{P} \subset \underline{K}(C(X \times S^1))$ be a finite subset which are required by Theorem 4.6 of [**33**] associated with $\epsilon/16$, $\mathcal{F}'$ and σ above. Without loss of generality, we may assume (by choosing smaller δ_1) that $\mathcal{G}_1 = \mathcal{F}_1 \otimes S$, where $\mathcal{F}_1 \subset C(X)$ is a finite subset of $C(X)$ and $S = \{z, 1_{C(S^1)}\}$. We may also assume that $\mathcal{F}_1 \supset \mathcal{F}$ and $\delta_1 < \epsilon/32$.

Let $\mathcal{P}_{C(X)} \subset \underline{K}(C(X))$ be associated with the Bott map as defined in 2.10. We may assume that $\beta(\mathcal{P}_{C(X)}) \subset \mathcal{P}$.

Let $\delta > 0$ and $\mathcal{G} \subset C(X)$ be a finite subset required by Lemma 6.4 associated with δ_1 (in place of ϵ), $\mathcal{F}_1$ (in place of $\mathcal{F}$), η, Δ (in placed of σ_1), δ_1 (in place of δ_0) and $\mathcal{G}_1$ (in place of $\mathcal{G}_0$).

Let h be a homomorphism and $u \in A$ be a unitary in the theorem associated with the above δ, $\mathcal{G}$. Since $\tau \circ h$ is $(\epsilon, \mathcal{F}, \mathcal{D}, \Delta)$-distributed, there exists an $\eta/2$-dense subset $\mathcal{D} = \{x_1, x_2, ..., x_m\}$ of X such that $O_i \cap O_j = \emptyset$, if $i \neq j$, where

$$O_i = \{x \in X : \mathrm{dist}(x, x_i) < \eta/2s\}, \; i = 1, 2, ..., m$$

and such that

$$\mu_{\tau \circ h}(O_i) \geq \eta\Delta, \quad i = 1, 2, ..., m \tag{e 7.10}$$

for all $\tau \in T(A)$.

By Lemma 6.4, there is a δ_1-$\mathcal{G}_1$-multiplicative contractive completely positive linear map $\phi : C(X \times S^1) \to A$ and a rectifiable continuous path of unitaries $\{w_t : t \in [0,1]\}$ of A such that

$$w_0 = u, \ \ \|[\phi(a), w_t]\| < \delta_1 \ \text{ for all } \ a \in \mathcal{G}_1, \tag{e 7.11}$$

$$\mathrm{Length}(\{w_t\}) \le \pi + \epsilon/4\pi, \tag{e 7.12}$$

$$\|h(a) - \phi(a \otimes 1)\| < \delta_1, \ \ \|\phi(a \otimes z) - h(a)w_1\| < \delta_1 \ \text{ for all } a \in \mathcal{F}_1 \ \text{ and} \tag{e 7.13}$$

$$\mu_{\tau\circ\phi}(O(x_i \times t_j)) > \frac{\Delta\eta}{2l} \ \text{ for all } \ \tau \in T(A), \tag{e 7.14}$$

where

$$O(x_i \times t_j) = \{(x \times t) \in X \times S^1 : \mathrm{dist}(x, x_i) < \eta/2s \ \text{ and } \ \mathrm{dist}(t, t_j) < \pi/4l\},$$

$i = 1, 2, ..., m, j = 1, 2, ..., l$ and $\{t_1, t_2, ..., t_l\}$ divides the unit circle into l arcs with the same length. We note that, by (e 7.13),

$$\|w_1 - \phi(1 \otimes z)\| < \delta_1. \tag{e 7.15}$$

Put $O_{i,j} = O(x_i \times t_j)$, $i = 1, 2, ..., m$ and $j = 1, 2, ..., l$. Then $O_{i,j} \cap O_{i',j'} = \emptyset$ if $(i,j) \ne (i',j')$. We also know that $\{x_i \times t_j : i = 1, 2, ..., m, j = 1, 2, ..., l\}$ is $\eta/2$-dense in $X \times S^1$. Moreover,

$$\mu_{\tau\circ\phi}(O_{i,j}) > 4\sigma\eta, i = 1, 2, ..., m \ \text{ and } \ j = 1, 2, ..., l. \tag{e 7.16}$$

Let $e_1, e_2 \in A$ be two non-zero mutually orthogonal projections such that

$$\tau(e_i) < \min\{\gamma/4, \frac{\sigma}{4}\eta\} \ \text{ for all } \ \tau \in T(A), \ \ i = 1, 2. \tag{e 7.17}$$

By 4.2, there is a unital homomorphism $h_1' : C(X) \to e_1Ae_1$ such that

$$[h_1'|_{C_0(Y_X)}] = [h|_{C_0(Y_X)}] \ \text{ in } \ KK(C(X), A). \tag{e 7.18}$$

By 2.25, there is a unital monomorphism $\psi_1' : C(X \times S^1) \to e_2Ae_2$ such that

$$[\psi_1'|_{C_0(Y_X)}] = 0 \ \text{ in } \ KK(C(X), A), \tag{e 7.19}$$

since it factors through $C([0,1])$.

Define $\psi_1 : C(X \times S^1) \to (e_1 + e_2)A(e_1 + e_2)$ by $\psi_1(a \otimes g) = h_1'(a)g(1)e_1 \oplus \psi_1'(a \otimes g)$ for all $a \in C(X)$ and $g \in C(S^1)$.

By 4.4, there is a unital homomorphism $\psi_2 : C(X \times S^1) \to (1-e)A(1-e)$ with finite dimensional range such that

$$|\tau \circ \phi(a) - \tau \circ \psi_2(a)| < \gamma/2 \ \text{ and } \ \mu_{\tau\circ\psi_2}(O_{i,j}) > 3\sigma\eta \ \text{ for all } \ \tau \in T(A) \tag{e 7.20}$$

and for all $a \in \mathcal{G}_1$.

Define $\psi : C(X \times S^1) \to A$ by $\psi(a) = \psi_1(a) \oplus \psi_2(a)$ for all $a \in C(X)$. It follows that (see 2.10)

$$[\psi]|_{\mathcal{P}} = [\phi]|_{\mathcal{P}}. \tag{e 7.21}$$

By (e 7.20) and (e 7.17),

$$|\tau \circ \psi(a) - \tau \circ \phi(a)| < \gamma \ \text{ for all } \ \tau \in T(A) \tag{e 7.22}$$

and for all $a \in \mathcal{G}_1$, and

$$\mu_{\tau\circ\psi}(O_{i,j}) > \sigma\eta \ \text{ for all } \ \tau \in T(A), \tag{e 7.23}$$

$i = 1, 2, ..., m$ and $j = 1, 2, ..., l$.

By Theorem 4.6 of [**33**] and the choices of δ_1 and $\mathcal{G}_1$, we obtain a unitary $Z \in A$ such that

$$\text{ad}\, Z \circ \psi \approx_{\epsilon/16} \phi \text{ on } \mathcal{F}'. \tag{e 7.24}$$

Note that $\text{ad}\, Z \circ \psi$ is a unital monomorphism. Put $H_1 = \text{ad}\, Z \circ \psi$ and $h_2 : C(X) \to A$ by $h_2(f) = H_1(f \otimes 1)$ for all $f \in C(X)$. So h_2 is a monomorphism. By (e 7.21) and $\boldsymbol{\beta}(\mathcal{P}_{C(X)}) \subset \mathcal{P}$, we conclude that

$$\text{Bott}(h_2(f), H_1(1 \otimes z)) = 0. \tag{e 7.25}$$

We now apply 6.1 to the monomorphism h_2 (for $f \in C(X)$) and the unitary $u' = H_1(1 \otimes z)$. We obtain a continuous path of unitaries $\{v_t : t \in [0,1]\} \subset A$ such that

$$v_0 = H_1(1 \otimes z), \;\; v_1 = 1_A, \;\; \|[\psi(f \otimes 1), v_t]\| < \epsilon/16 \tag{e 7.26}$$

for all $t \in [0,1]$ and for all $f \in \mathcal{F}$. Moreover,

$$\text{Length}(\{v_t\}) \le \pi + \epsilon\pi/8. \tag{e 7.27}$$

By (e 7.24) and (e 7.13),

$$\|[h(a), v_t]\| < \epsilon/8 \text{ for all } a \in \mathcal{F} \text{ and for all } t \in [0,1]. \tag{e 7.28}$$

Then by (e 7.15)

$$\|v_1 - w_1\| < \epsilon/8. \tag{e 7.29}$$

By connecting the path $\{w_t\}$ and $\{v_t\}$ appropriately, we see that the theorem follows. □

The following follows immediately from 7.4 and the discussion below 7.3. Note that $\eta = \eta_d(\epsilon, \mathcal{F}$ is determined by ϵ and $\mathcal{F}$.

Corollary 7.5. *Let X be a connected finite CW-complex with a fixed metric and let $\Delta : (0,1) \to (0,1)$ be an increasing map. For any $\epsilon > 0$, any finite subset $\mathcal{F} \subset C(X)$, a unital separable simple C^*-algebra A with tracial rank zero and a unital monomorphism $h : C(X) \to A$ for which $\mu_{\tau \circ h}$ is Δ-distributed for all $\tau \in T(A)$, there exist $\delta > 0$ and a finite subset $\mathcal{G} \subset C(X)$ satisfying the following: If $u \in A$ is a unitary such that*

$$\|[h(a), u]\| < \delta \textit{ for all } a \in \mathcal{G} \textit{ and } \text{Bott}(h,u) = 0. \tag{e 7.30}$$

Then, there exists a continuous rectifiable path of unitaries $\{u_t : t \in [0,1]\}$ of A such that

$$u_0 = u, \;\; u_1 = 1_A \textit{ and } \|[h(a), u_t]\| < \epsilon \textit{ for all } a \in \mathcal{F}, t \in [0,1]. \tag{e 7.31}$$

Moreover,

$$\text{Length}(\{u_t\}) \le 2\pi + \epsilon\pi. \tag{e 7.32}$$

Of course we can omit the mention of Δ, if we allows δ depends on h.

Corollary 7.6. *Let X be a finite CW-complex with a fixed metric. For any $\epsilon > 0$, any finite subset $\mathcal{F} \subset C(X)$, a unital separable simple C^*-algebra A with tracial rank zero and a unital monomorphism $h : C(X) \to A$, there exist $\delta > 0$ and a finite subset $\mathcal{G} \subset C(X)$ satisfying the following: If $u \in A$ is a unitary such that*

$$\|[h(a), u]\| < \delta \textit{ for all } a \in \mathcal{G} \textit{ and } \text{Bott}(h,u) = 0.$$

Then, there exists a continuous rectifiable path of unitaries $\{u_t : t \in [0,1]\}$ of A such that

$$u_0 = u,\ \ u_1 = 1_A \ \text{and}\ \|[h(a), u_t]\| < \epsilon \ \text{for all}\ \ a \in \mathcal{F} \ \text{and}\ t \in [0,1].$$

Moreover,

$$\text{Length}(\{u_t\}) \le 2\pi + \epsilon.$$

8. The Basic Homotopy Lemma — compact metric spaces

Theorem 8.1. *Let X be a compact metric space which is a compact subset of a finite CW complex. For any $\epsilon > 0$, any finite subset $\mathcal{F} \subset C(X)$ and any increasing map $\Delta : (0,1) \to (0,1)$, there exists $\delta > 0$, a finite subset $\mathcal{G} \subset C(X)$ and a finite subset $\mathcal{P} \subset \underline{K}(C(X))$ satisfying the following:*

Suppose that A is a unital separable simple C^*-algebra with tracial rank zero, $h : C(X) \to A$ is a unital monomorphism and $u \in A$ is a unitary such that $\mu_{\tau\circ h}$ is Δ-distributed for all $\tau \in T(A)$,

$$\|[h(a), u]\| < \delta \ \text{for all}\ \ a \in \mathcal{G} \ \text{and}\ \text{Bott}(h, u)|_{\mathcal{P}} = 0. \tag{e 8.1}$$

Then, there exists a continuous rectifiable path of unitaries $\{u_t : t \in [0,1]\}$ of A such that

$$u_0 = u,\ \ u_1 = 1_A \ \text{and}\ \|[h(a), u_t]\| < \epsilon \ \text{for all}\ \ a \in \mathcal{F} \ \text{and}\ t \in [0,1] \tag{e 8.2}$$

Moreover,

$$\text{Length}(\{u_t\}) \le 2\pi + \epsilon\pi \tag{e 8.3}$$

Proof. Without of loss of generality, we may assume that there is a finite CW complex Y' such that $X \subset Y'$ is a compact subset (with a fixed metric).

Let $\epsilon > 0$ and $\mathcal{F} \subset C(X)$ be a finite subset. We may assume that $1 \in \mathcal{F}$. Let $S = \{1_{C(S^1)}, z\}$. To simplify notation, without loss of generality, we may assume that $\mathcal{F}$ is a subset of the unit ball of $C(X)$.

Let

$$\mathcal{F}' = \{f \otimes a : f \in \mathcal{F} \ \text{and}\ a \in S\} \subset C(X \times S^1).$$

Let $\eta_1 > 0$ so that

$$\eta_1 < \min\{\frac{\epsilon}{64(1 + \max\{\|f\| : f \in \mathcal{F}\})}, \sigma_{X,\mathcal{F},\epsilon/64}\}.$$

It follows that if $g \in \mathcal{F}'$, then

$$|g(z) - g(z')| < \epsilon/64 \ \text{for all}\ \ g \in \mathcal{F}'$$

if $\text{dist}(z, z') < \eta$ $(z, z' \in X \times S^1)$.

By the Tietze Extension Theorem, one has $g_f \in C(Y')$ such that $(g_f)|_X = f$ and $\|g_f\| = \|f\|$ for all $f \in \mathcal{F}$. Put $\mathcal{G}'_{00} = \{g_f \in C(Y') : f \in \mathcal{F}\}$. Let $\eta_0 = \sigma_{Y',\mathcal{G}'_{00},\epsilon/64}$. It is easy to construct a finite CW complex $Y \subset Y'$ such that $X \subset Y$ and X is $\min\{\eta_0/2, \eta_1/4\}$-dense in Y. Put

$$\mathcal{G}_0 = \{g \in C(Y) : g|_Y = g_f \ \text{for some}\ \ f \in \mathcal{F}\}.$$

We estimate that

$$\sigma_{Y,\mathcal{G}_0,\epsilon/32} \ge \sigma_{X,\mathcal{F},\epsilon/64}. \tag{e 8.4}$$

Let $\eta = \eta_1/2$. Put

$$\mathcal{G}'_0 = \{g \otimes f \in C(Y \times S^1) : f \in \mathcal{G}_0 \ \text{and}\ g \in S\}.$$

It follows that if $g \in \mathcal{G}_0'$, then

$$|g(z) - g(z')| < \epsilon/32 \text{ for all } g \in \mathcal{G}_0'$$

if $\mathrm{dist}(z, z') < \eta$ $(z, z' \in Y \times S^1)$.

Suppose that X contains an $\eta/2$-dense subset which is (s, η)-separated. Let $\Delta : (0,1) \to (0,1)$ be an increasing map and $\Delta_0 = \Delta(\eta/2s)/2s$.

Choose an integer l such that $256\pi/l < \min\{\eta/4s, \epsilon/4\}$. Let

$$\sigma = \Delta_0/8l.$$

Let $\gamma > 0$, $\delta_1 > 0$, $\mathcal{G}_1 \subset C(Y \times S^1)$ be a finite subset and $\mathcal{P}_1 \subset \underline{K}(C(Y \times S^1))$ be a finite subset which are required by Theorem 4.6 of [**33**] associated with $\epsilon/4$, $\mathcal{G}_0'$ and σ above. Without loss of generality, we may assume (by choosing smaller δ_1) that $\mathcal{G}_1 = \mathcal{F}_1 \otimes S$, where $\mathcal{F}_1 \subset C(Y)$ is a finite subset of $C(Y)$. We may also assume that $\mathcal{F}_1 \supset \mathcal{F}$ and $\delta_1 < \epsilon/4$.

Let $\delta > 0$ and $\mathcal{G} \subset C(X)$ be a finite subset required by Lemma 6.4 associated with δ_1 (in place of ϵ), $\mathcal{F}_1$ (in place of $\mathcal{F}$), η, Δ_0 (in place of σ_1) and with δ_1 (in place of δ_0) and $\mathcal{G}_1$ (in place of $\mathcal{G}_0$).

Let $\mathcal{P} = [\theta](\mathcal{P}_1)$, where $\theta : C(Y) \to C(X)$ is the quotient map. Let h be a homomorphism and $u \in A$ be a unitary in the theorem associated with the above δ, $\mathcal{G}$ and $\mathcal{P}$. Moreover, we assume that $\tau \circ h$ is $(\epsilon, \mathcal{F}, \mathcal{D}, \Delta_0)$-distributed, so there exists an $\eta/2$-dense subset $\mathcal{D} = \{x_1, x_2, ..., x_m\}$ of X such that $O_i \cap O_j = \emptyset$, if $i \neq j$, where

$$O_i = \{x \in X : \mathrm{dist}(x, x_i) < \eta/2s\}, \quad i = 1, 2, ..., m$$

for $s = s(\eta) > 0$ given above and such that

$$\mu_{\tau \circ h}(O_i) \geq \eta \Delta_0, \quad i = 1, 2, ..., m \tag{e 8.5}$$

for all $\tau \in T(A)$. Note here we assume that $\eta/s < \min\{\mathrm{dist}(x_i, x_j) : i \neq j\}$.

Note also that $\{x_1, x_2, ..., x_m\}$ is $\eta_1/2$-dense in Y.

By Lemma 6.4, there is a $\mathcal{G}_1$-δ_1-multiplicative contractive completely positive linear map $\phi : C(X \times S^1) \to A$ and a rectifiable continuous path of unitaries $\{w_t : t \in [0,1]\}$ of A such that

$$w_0 = u, \quad \|[\phi(a), w_t]\| < \delta_1 \text{ for all } a \in \mathcal{G}_1, \tag{e 8.6}$$

$$\mathrm{Length}(\{w_t\}) \leq \pi + \epsilon \cdot \pi/4, \tag{e 8.7}$$

$$\|h(a) - \phi(a \otimes 1)\| < \delta_1, \quad \|\phi(a \otimes z) - h(a)w_1\| < \delta_1 \text{ for all } a \in \mathcal{F}_1 \text{ and} \tag{e 8.8}$$

$$\mu_{\tau \circ \phi}(O(x_i \times t_j)) > \frac{\Delta_0 \eta}{2l} \text{ for all } \tau \in T(A), \tag{e 8.9}$$

where

$$O(x_i \times t_j) = \{(x \times t) \in X \times S^1 : \mathrm{dist}(x, x_i) < \eta/2s \text{ and } \mathrm{dist}(t, t_j) < \pi/4sl\},$$

$i = 1, 2, ..., m, j = 1, 2, ..., l$ and $\{t_1, t_2, ..., t_l\}$ divides the unit circle into l arcs with the same length.

Put

$$\tilde{O}_{i,j} = \{(y \times t) \in Y \times S^1 : \mathrm{dist}(y, x_i) < \eta/2s \text{ and } \mathrm{dist}(t, t_j) < \pi/4sl\},$$

$i = 1, 2, ..., m$ and $j = 1, 2, ..., l$. Then $\tilde{O}_{i,j} \cap \tilde{O}_{i',j'} = \emptyset$ if $(i,j) \neq (i',j')$. We also know that $\{x_i \times t_j : i = 1, 2, ..., m, j = 1, 2, ..., l\}$ is $\eta/2$-dense in $X \times S^1$. Define $\Phi : C(Y \times S^1) \to A$ by $\Phi(f) = \phi(\theta(f))$ for all $f \in C(Y \times S^1)$. Moreover,

$$\mu_{\tau \circ \Phi}(\tilde{O}_{i,j}) > 4\sigma\eta, i = 1, 2, ..., m \text{ and } j = 1, 2, ..., l. \tag{e 8.10}$$

Write $Y = \sqcup_{k=1}^R Y_k$, where each Y_k is a connected finite CW complex. Fix a point $\xi_k \in Y_k$ for each k. Let E_k be the characteristic function for Y_k, $k = 1, 2, ..., R$. For convenience, one may assume that $[E_k] \in \mathcal{P}_1$. Denote by $E_k' = E_k \times 1_{C(S^1)}$, $k = 1, 2, ..., R$.

Put $e_k = h(\theta(E_k))$, $k = 1, 2, ..., R$. Choose nonzero projections $d_k \le e_k$ in A, $k = 1, 2, ..., R$ such that

$$\tau(d_k) < \min\{\frac{\gamma}{4R}, \frac{\sigma\cdot\eta}{4R}\} \text{ for all } \tau \in T(A), \tag{e 8.11}$$

$k = 1, 2, ..., R$. We may assume that $e_k - d_k \neq 0$, $k = 1, 2, ..., R$.

By 4.4, there is a unital homomorphism $\psi_{2,k} : C(Y_i \times S^1) \to (e_k - d_k)A(e_k - d_k)$ with finite dimensional range such that

$$|\tau\circ(\phi(E_k' a E_k')) - \tau\circ\psi_{2,k}(a)| < \min\{\gamma/2R, \frac{\sigma\cdot\eta}{2R}\} \text{ for all } \tau \in T(A) \tag{e 8.12}$$

and for all $a \in \mathcal{G}_1$, $k = 1, 2, ..., R$. Set $e = \sum_{i=1}^R d_i$. Then, by (e 8.11),

$$\tau(e) < \min\{\frac{\gamma}{4}, \frac{\sigma\cdot\eta}{4}\} \text{ for all } \tau \in T(A). \tag{e 8.13}$$

Define $\psi_2 : C(Y \times S^1) \to (1-e)A(1-e)$ by $\psi_2(f) = \sum_{k=1}^R \psi_{2,k}(E_k' f)$ for $f \in C(Y \times S^1)$. Then we have

$$|\tau\circ(\phi\circ\theta(a)) - \tau\circ\psi_2(a)| < \min\{\gamma/2, \frac{\sigma\cdot\eta}{2}\} \text{ and } \mu_{\tau\circ\psi_2}(\tilde{O}_{i,j}) > 3\sigma\eta \tag{e 8.14}$$

for all $\tau \in T(A)$ and for all $a \in \mathcal{G}_1$.

By 4.2, there is a unital monomorphism $h_{1,k} : C(Y) \to d_k A d_k$ such that

$$[h_{1,k}|_{C_0(Y_k\setminus\{\xi_k\})}] = [(h\circ\theta)|_{C_0(Y_k\setminus\{\xi_k\})}] \text{ in } KL(C(Y_k), A), \tag{e 8.15}$$

$k = 1, 2, ..., R$. Define $h_1(f) = \oplus_{k=1}^R h_{1,k}(E_k f)$ for $f \in C(Y)$.

Define $\psi_1 : C(Y \times S^1) \to eAe$ by $\psi_1(a \otimes g) = h_1(a)g(1)e$ for all $a \in C(Y)$ and $g \in C(S^1)$.

Define $\psi : C(Y \times S^1) \to A$ by $\psi(a) = \psi_1(a) \oplus \psi_2(a)$ for all $a \in C(Y)$. We have $h_1(E_k) = e_k = h\circ\theta(E_k)$, $k = 1, 2, ..., R$. It follows from this, (e 8.15) and 4.1 that

$$[\psi]|_{\mathcal{P}_1} = [\phi\circ\theta]|_{\mathcal{P}_1} \tag{e 8.16}$$

By (e 8.14) and (e 8.13),

$$|\tau\circ\psi(a) - \tau\circ\phi\circ\theta(a)| < \gamma \text{ for all } \tau \in T(A) \tag{e 8.17}$$

and for all $a \in \mathcal{G}_1$, and

$$\mu_{\tau\circ\psi}(\tilde{O}_{i,j}) > \sigma\eta \text{ for all } \tau \in T(A), \tag{e 8.18}$$

$i = 1, 2, ..., m$ and $j = 1, 2, ..., l$.

By Theorem 4.6 of [**33**] and the choices of δ_1 and $\mathcal{G}_1$, we obtain a unitary $Z \in A$ such that

$$\operatorname{ad} Z\circ\psi \approx_{\epsilon/4} \phi\circ\theta \text{ on } \mathcal{G}_0'. \tag{e 8.19}$$

Since ψ_2 has finite dimensional range, from the definition of ψ_1 and ψ, we obtain a rectifiable continuous path of unitaries $\{U_t : t \in [0,1]\}$ of A with

$$\text{Length}(\{U_t\}) \le \pi + \epsilon\pi/2 \tag{e 8.20}$$

such that

$$U_0 = \psi(1\otimes z),\ U_1 = 1_A \text{ and } \|[\psi(a\otimes 1), U_t]\| = 0 \tag{e 8.21}$$

for all $a \in \mathcal{F}$ and for all $t \in [0,1]$.

Put $W_t = Z^*U_tZ$ for $t \in [0,1]$. Then $W_1 = 1_A$. We also have that (by (e 8.8) and (e 7.24))

$$\|W_0 - u\| < \delta_1 + \|W_0 - \phi(1 \otimes z)\| < \epsilon/4 + \epsilon/4 = \epsilon/2. \tag{e 8.22}$$

Moreover,

$$\|[W_t, \phi(a \otimes 1)\| < \epsilon/2 \text{ for all } a \in \mathcal{F}. \tag{e 8.23}$$

By (e 8.8), (e 8.19) and using the fact (e 8.22) and (e 8.20), and using the path $\{w_t\}$ and the path $\{W_t : t \in [0,1]\}$, we obtain a rectifiable continuous path of unitaries $\{u_t : t \in [0,1]\}$ of A such that

$$u_0 = u, \;\; u_1 = 1_A, \;\; \|[h(a), u_t]\| < \epsilon \tag{e 8.24}$$

for all $a \in \mathcal{F}$ and for all $t \in [0,1]$ and

$$\text{Length}(\{u_t\}) \le 2\pi + \epsilon\pi. \tag{e 8.25}$$

□

Remark 8.2. One can absorb the proof of Theorem 7.4 into that of Theorem 8.1. However, it does not seem necessarily helpful to mix the issue of dependence of δ on a local measure distribution $(X, \mathcal{F}, \mathcal{D}, \Delta)$ together with other constants such as η and s with the issue of reduction of a compact subset X of a finite CW complex to the special situation that X is a finite CW complex.

Theorem 8.3. *Let X be a compact metric space. For any $\epsilon > 0$, any finite subset $\mathcal{F} \subset C(X)$ and any increasing map $\Delta : (0,1) \to (0,1)$, there exists $\delta > 0$, a finite subset $\mathcal{G} \subset C(X)$ and a finite subset $\mathcal{P} \subset \underline{K}(C(X))$ satisfying the following:*

Suppose that A is a unital separable simple C^*-algebra with tracial rank zero, $h : C(X) \to A$ is a unital monomorphism and $u \in A$ is a unitary such that $\mu_{\tau\circ h}$ is Δ-distributed for all $\tau \in T(A)$,

$$\|[h(a), u]\| < \delta \text{ for all } a \in \mathcal{G} \text{ and } \text{Bott}(h, u)|_{\mathcal{P}} = 0. \tag{e 8.26}$$

Then, there exists a continuous rectifiable path of unitaries $\{u_t : t \in [0,1]\}$ of A such that

$$u_0 = u, \;\; u_1 = 1_A \text{ and } \|[h(a), u_t]\| < \epsilon \text{ for all } a \in \mathcal{F} \text{ and } t \in [0,1]. \tag{e 8.27}$$

Moreover,

$$\text{Length}(\{u_t\}) \le 2\pi + \epsilon\pi. \tag{e 8.28}$$

Proof. It is well known (see, for example, [**44**]) that there exists a sequence of finite CW complex Y_n such that $C(X) = \lim_{n\to\infty}(C(Y_n), \lambda_n)$. By replacing Y_n by one of its compact subset X_n, we may write $C(X) = \lim_{n\to\infty}(C(X_n), \lambda_n)$, where each λ_n is a monomorphism form $C(X_n)$ into $C(X)$. Thus there is a surjective homeomorphism $\alpha_n : X \to X_n$ such that $\lambda_n(f) = f \circ \alpha_n$ for all $f \in C(X_n)$.

Now fix an $\epsilon > 0$ and a finite subset $\mathcal{F} \subset C(X)$. Let $\eta_1 = \sigma_{X,\mathcal{F},\epsilon/64}$.

There is an integer $N \ge 1$ and a finite subset $\mathcal{F}_1 \subset C(X_N)$ such that, for each $f \in \mathcal{F}$, there exists $g_f \in C(X_N)$ such that

$$\|f - \lambda_N(g_f)\| < \epsilon/4. \tag{e 8.29}$$

Suppose that $\Delta : (0,1) \to (0,1)$ is an increasing map. For each $1 > \eta > 0$ and $y \in X_N$, there is $x(y) \in X$ and $1 > r(y,\eta) > 0$ such that

$$\alpha_N(O(x(y), r(y,\eta))) \subset O(y,\eta). \tag{e 8.30}$$

Since

$$\cup_{x\in X_N} O(x, \eta/2) \supset X_N,$$

there are $y_1, y_2, ..., y_m \in X_N$ such that

$$\cup_{k=1}^m O(y_i, \eta/2) \supset X_N.$$

Now for any $y \in X_N$, $y \in O(y_k, \eta/2)$ for some $k \in \{1,2,...,m\}$. Then

$$O(y,\eta) \supset O(y_k, \eta/2).$$

Put

$$\bar{\eta} = \min\{r(y_k, \eta) : k = 1,2,...,m\} > 0$$

Then, for any $y \in X_N$, there exists $x(y) \in X$ such that

$$\alpha_N(O(x(y), \bar{\eta})) \subset O(y,\eta). \tag{e 8.31}$$

Fix $1 > \eta > 0$. Denote by $\bar{\bar{\eta}}$ the supremum of such $\bar{\eta}$ for which (e 8.31) holds for all $y \in X_N$. Then $\bar{\bar{\eta}}_1 \le \bar{\bar{\eta}}_2$, if $0 < \eta_1 < \eta_2 < 1$. Moreover, for any $0 < \eta' < \bar{\bar{\eta}}$ and any $y \in X_N$, there exists $x(y) \in X$ such that

$$\alpha_N(O(x(y), \eta')) \subset O(y,\eta). \tag{e 8.32}$$

Define $\Delta_1 : (0,1) \to (0,1)$ by

$$\Delta_1(\eta) = \Delta(\bar{\bar{\eta}}/2) \cdot \frac{\bar{\bar{\eta}}}{2} \text{ for all } \eta \in (0,1).$$

Note that Δ_1 is increasing since Δ is. Let $\delta > 0$, $\mathcal{G}_1 \subset C(X_N)$ be a finite subset and $\mathcal{P}_1 \subset \underline{K}(C(X_N))$ be a finite subset required by 8.1 associated with $\epsilon/4$ (in place of ϵ), $\lambda_N(\mathcal{F}_1)$ (in place of $\mathcal{F}$) and $\Delta_1 : (0,1) \to (0,1)$.

Put $\mathcal{P} = [(\lambda_N)](\mathcal{P}_1)$ in $\underline{K}(C(X))$.

Suppose that $h : C(X) \to A$ is a unital monomorphism and $u \in A$ is a unitary satisfying the conditions in the theorem associated with the above δ, $\mathcal{G}$, $\mathcal{P}$.

Define $h_1 : C(X_N) \to A$ by $h_1 = h \circ \lambda_N$. Then

$$\|[h_1(g), u]\| < \delta \text{ for all } g \in \mathcal{G}_1,$$
$$\text{Bott}(h_1, u)|_{\mathcal{P}_1} = 0.$$

Moreover one checks that $\mu_{\tau\circ h_1}$ is Δ_1-distributed. To see this, let $1 > \eta > 0$ and let $O(y,\eta) \subset X_N$ be an open ball with radius η. Then, by (e 8.32), there exists $x(y) \in X$ such that

$$\alpha_N(O(x(y), \bar{\bar{\eta}}/2)) \subset O(y,\eta).$$

It follows that, for any $1 > \eta > 0$,

$$\mu_{\tau\circ h_1}(O(y,\eta)) = \mu_{\tau\circ h}(\alpha_N^{-1}(O(y,\eta))) \ge \mu_{\tau\circ h}(O(x(y), \bar{\bar{\eta}}/2)) \tag{e 8.33}$$
$$\ge \Delta(\bar{\bar{\eta}}/2) \cdot (\bar{\bar{\eta}}/2\eta) \cdot \eta \ge \Delta(\bar{\bar{\eta}}/2) \cdot (\bar{\bar{\eta}}/2) \cdot \eta = \Delta_1(\eta)\eta. \tag{e 8.34}$$

This shows that $\mu_{\tau\circ h_1}$ is Δ_1-distributed for all $\tau \in T(A)$. By the choices of δ, $\mathcal{G}_1$ and $\mathcal{P}_1$, by applying 8.1, one obtains a continuous path of unitaries $\{u_t : t \in [0,1]\} \subset A$ with

$$\text{Length}(\{u_t\}) \le 2\pi + \epsilon\pi$$

such that

$$u_0 = u,\ \ u_1 = 1 \text{ and } \|[h_1(f), u_t]\| < \epsilon/4 \tag{e 8.35}$$

for all $f \in \mathcal{F}_1$. It follows from (e 8.35) and (e 8.29) that

$$\|[h(f), u_t]\| < \epsilon \text{ for all } f \in \mathcal{F}.$$

□

Corollary 8.4. *Let X be a compact metric space. For any $\epsilon > 0$, any finite subset $\mathcal{F} \subset C(X)$, a unital separable simple C^*-algebra A with tracial rank zero and a unital monomorphism $h : C(X) \to A$, there exists $\delta > 0$, a finite subset $\mathcal{G} \subset C(X)$ and a finite subset $\mathcal{P} \subset \underline{K}(C(X))$ satisfying the following: If $u \in A$ is a unitary such that*

$$\|[h(a), u]\| < \delta \textit{ for all } a \in \mathcal{G} \textit{ and } \mathrm{Bott}(h, u)|_{\mathcal{P}} = 0, \tag{e 8.36}$$

then, there exists a continuous rectifiable path of unitaries $\{u_t : t \in [0,1]\}$ of A such that

$$\|[h(a), u_t]\| < \epsilon \textit{ for all } a \in \mathcal{F}, u_0 = u \textit{ and } u_1 = 1_A. \tag{e 8.37}$$

Moreover,

$$\mathrm{Length}(\{u_t\}) \le 2\pi + \epsilon. \tag{e 8.38}$$

Corollary 8.5. *Let X be a compact metric space. Let $\epsilon > 0$, let $\mathcal{F} \subset C(X)$ be a finite subset, let A be a unital simple separable simple C^*-algebra with tracial rank zero and let $h : C(X) \to A$ be a unital monomorphism.*

If there is a unitary $u \in A$ such that

$$h(a)u = uh(a) \textit{ for all } a \in C(X) \textit{ and } \mathrm{Bott}(h, u) = 0. \tag{e 8.39}$$

Then, there exists a continuous rectifiable path of unitaries $\{u_t : t \in [0,1]\}$ of A such that

$$\|[h(a), u_t]\| < \epsilon \textit{ for all } a \in \mathcal{F}, u_0 = u \textit{ and } u_1 = 1_A. \tag{e 8.40}$$

Moreover,

$$\mathrm{Length}(\{u_t\}) \le 2\pi + \epsilon. \tag{e 8.41}$$

9. The constant δ and an obstruction behind the measure distribution

The reader undoubtedly notices that the constant δ in Theorem 3.7 and in 3.13 is universal. But the constant δ in Theorem 7.4 and in Theorem 8.3 is not quite universal, it depends on the measure distribution $\mu_{\tau\circ h}$. The introduction of the concept of measure distribution is not just for the convenience of the proof. In fact it is essential. In this section we will demonstrate that when $\dim(X) \ge 2$, the constant can not be universal for any such space X whenever A is a separable non-elementary simple C^*-algebra with real rank zero and stable rank one. There is a topological obstruction for the choice of δ hidden behind the measure distribution.

Lemma 9.1. *Let A be a unital separable simple C^*-algebra with stable rank one. Suppose that $\{d_n\}$ and $\{e_n\}$ are two sequences of non-zero projections such that there are $k(n)$ mutually orthogonal and mutually equivalent projections $q(n,1)$, $q(n,2)$, ..., $q(n, k(n))$ in e_nAe_n each of which is equivalent to d_n, $n = 1, 2, ...$*

Let $d = \{d_n\} \in l^\infty(A)$ and let I_d be the (closed two-sided) ideal of $l^\infty(A)$ generated by d. Then $e = \{e_n\} \notin I_d$ if $lim_{k\to\infty}k(n) = \infty$. Moreover, $\pi(e) \notin \pi(I_d)$, where $\pi : l^\infty(A) \to q_\infty(A)$ is the quotient map.

PROOF. Assume that $\pi(e) \in \pi(I_d)$. Then, by Lemma 3.3.6 of [**29**], there are $x_1, x_2, ..., x_m \in l^\infty(A)$ such that

$$\sum_{i=1}^{m} \pi(x_i^*)\pi(d)\pi(x_i) = \pi(e). \tag{e 9.1}$$

Put $\tilde{e} = \mathrm{diag}(e, \overbrace{0, \cdots, 0}^{m})$ and $\tilde{d} = \mathrm{diag}(\overbrace{d, d, \cdots, d}^{m})$. Then it is standard (see the proof of Proposition of 3.3.7 of [**29**], for example) that there is a projection $\bar{g} \le \pi(\tilde{d})$ and a partial isometry $\bar{v} \in M_m(l^\infty(A)/c_0(A))$ such that

$$\bar{v}^*\bar{v} = \pi(\tilde{e}) \text{ and } \bar{v}\bar{v}^* = \bar{g}.$$

Choose a projection $g = \{g_n\} \in M_m(l^\infty(A))$ and $v = \{y_n\} \in M_m(l^\infty(A))$ such that $\pi(g) = \bar{g}$ and $\pi(v) = \bar{v}$. Then one has

$$\lim_{n\to\infty} \|y_n^* y_n - e_n\| = 0, \ \lim_{n\to\infty} \|y_n y_n^* - g_n\| = 0 \tag{e 9.2}$$

$$\text{and } \lim_{n\to\infty} \|g_n - \bar{d}_n g_n \bar{d}_n\| = 0, \tag{e 9.3}$$

where $\bar{d}_n = \mathrm{diag}(\overbrace{d_n, d_n, \cdots, d_n}^{m}) \in M_m(A)$, $n = 1, 2, ...$ Then, by a standard perturbation argument (see, for example, Lemma 2.5.3 and Lemma 2.5.5 of [**29**]), one obtains a sequence of partial isometries $\{v_n\} \subset M_m(A)$ such that, for all $n \ge N$ (for some integer $N \ge 1$),

$$v_n^* v_n = e_n \text{ and } v_n v_n^* \le \mathrm{diag}(\overbrace{d_n, d_n \cdots, d_n}^{m}). \tag{e 9.4}$$

Let τ be a quasitrace of A (given by [**2**]). It follows from (e 9.4) that

$$k(n)\tau(d_n) \le \tau(e_n) \le m\tau(d_n) \text{ for all } n \ge N.$$

This is not possible. □

Lemma 9.2. *Let X be a connected finite CW complex with dimension at least 2 and let $Y \subset X$ be a compact subset which is homeomorphic to a k-cell with $k \ge 2$. Let A be a unital C^*-algebra and let $h_1 : C(Y) \to A$ be a unital homomorphism. Define $h = h_1 \circ \pi$, where $\pi : C(X) \to C(Y)$ is the quotient map. Then, for any sufficiently small δ, any sufficiently large finite subset $\mathcal{G} \subset C(X)$ and any unitary $u \in A$ with $[u] = 0$,*

$$\mathrm{Bott}(h, u) = 0, \tag{e 9.5}$$

provided that $\|[h(a), u]\| < \delta$ for all $a \in \mathcal{G}$.

PROOF. Let $\{p_i \in M_{n(i)}(C(X)) : i = 1, 2, ..., k\}$ be k projections which generate $K_0(C(X))$. Assume that p_i has rank $r(i) \le n(i)$, $i = 1, 2, ..., k$. Since Y is contractive, there is a unitary $w_i \in M_{n(i)}(C(Y))$ such that

$$w_i^* \pi(p_i) w_i = 1_{r(i)}, \ i = 1, 2, ..., k, \tag{e 9.6}$$

where $1_{r(i)}$ is a diagonal matrix in $M_{n(i)}$ with $r(i)$ 1's and $n(i) - r(i)$ zeros. There is $a_i \in M_{n(i)}(C(X))$ such that $\pi(a_i) = w_i$, $i = 1, 2, ..., k$.

Put

$$\overline{u}^{n(i)} = \mathrm{diag}(\overbrace{u, u, ..., u}^{n(i)}).$$

Then, for sufficiently small δ and sufficiently large $\mathcal{G}$,

$$h_1(w_i^*)(h(p_i)\overline{u}^{n(i)})h_1(w_i) = h(a_ip_i)\overline{u}^{n(i)}h(a_i) \tag{e 9.7}$$

$$\approx_{1/16} 1_{r(i)}\overline{u}^{n(i)}. \tag{e 9.8}$$

It implies that

$$[h(p_i)\overline{u}^{n(i)}] = 0 \ \ (\text{in } K_1(A)). \tag{e 9.9}$$

This implies that

$$\mathrm{bott}_0(h, u) = 0.$$

The fact that

$$\mathrm{bott}_1(h, u) = 0$$

follows from the fact that $K_1(C(Y)) = \{0\}$.

Since $K_i(C(Y \times S^1))$ is torsion free ($i = 0, 1$), we conclude that

$$\mathrm{Bott}(h, u) = 0.$$

□

A continuous path with length L may have infinite Lipschitz constant. Note in the following lemma, $\{v_t : t \in [0, 1]\}$ not only satisfies the Lipschitz condition but also its length does not increase.

Lemma 9.3. *Let $L > 0$ be a positive number, let C be a unital C^*-algebra and let $M > 0$. Then, for any $\epsilon > 0$, there is $\delta > 0$ satisfying the following:*

Let A be a unital C^*-algebra, let $\mathcal{G} \subset C$ be a self-adjoint subset with $\|f\| \le M$ for all $f \in \mathcal{G}$, let $\{u_t : t \in [0, 1]\}$ be a continuous path of unitaries in A and let $h : C \to A$ be a monomorphism such that

$$u_1 = 1, \ \ \|[h(g), u_t]\| < \delta \ \textit{for all} \ g \in \mathcal{G} \ \textit{and} \ \ \textit{for all} \ t \in [0, 1]$$

and

$$\mathrm{Length}(\{u_t\}) \le L. \tag{e 9.10}$$

Then there is a path of unitaries $\{v_t : t \in [0, 1]\}$ in A such that

$$v_0 = u_0, \ \ v_1 = 1,$$
$$\|[h(g), v_t]\| < \epsilon \ \textit{for all} \ g \in \mathcal{G} \ \textit{and} \ t \in [0, 1],$$
$$\mathrm{Length}(\{v_t\}) \le L.$$

Moreover,

$$\|v_t - v_{t'}\| \le L|t - t'|$$

for all $t, t' \in [0, 1]$.

Proof. Let $0 = t_0 < t_1 < \cdots < t_m = 1$ be a partition of $[0, 1]$ such that

$$0 < \mathrm{Length}(\{u_t : t \in [t_{i-1}, t_i]\} \le \pi/2, \tag{e 9.11}$$

$i = 1, 2, ..., m$.

Put $w_i = u_{t_{i-1}}^* u_{t_i}$, $i = 1, 2, ..., m$. Then there is a rectifiable continuous path from w_i to 1 with length no more than $\pi/2$. It follows that

$$sp(w_i) \subset S = \{e^{i\pi t} : t \in [-1/2, 1/2]\}.$$

Then

$$F(e^{i\pi t}) = t$$

defines a continuous function from S to $[-1/2, 1/2]$. Then

$$w_i = \exp(\sqrt{-1}\,\pi F(w_i)), \quad i = 1, 2, ..., m.$$

The function F does not depend on w_i nor A.

Since $\mathcal{G}$ is self-adjoint and $\|f\| \le M$ for all $f \in \mathcal{G}$, there is a positive number δ depending on ϵ, M and F only, such that, if

$$\|[h(g), u_t]\| < \delta \text{ for all } g \in \mathcal{G} \text{ and } t \in [0,1],$$

then

$$\|[h(g), \exp(\sqrt{-1}\,\pi t\, F(w_i))]\| < \epsilon \tag{e 9.12}$$

for all $t \in [0,1]$ (cf. 2.5.11 of [**29**] and 2.6.10 of [**29**]). Let

$$l_i = \text{Length}(\{u_t : t \in [t_{i-1}, t_i]\}), \quad i = 1, 2, ..., m.$$

Then

$$\|\pi F(w_i)\| = l_i, \quad i = 1, 2, ..., m.$$

Now, define

$$v_0 = u_0, \;\; v_t = u_{i-1}(exp(\sqrt{-1}\pi \frac{t - s_{i-1}}{s_i - s_{i-1}} F(w_i))) \text{ for all } t \in [s_{i-1}, s_i],$$

where $s_0 = 0$, $s_i = \sum_{j=1}^{i} l_j/L$, $i = 1, 2, ..., m$. Clearly, if $t, t' \in [s_{i-1}, s_i]$, then

$$\begin{aligned} \|v_t - v_{t'}\| &= \|\exp(\sqrt{-1}\pi \frac{t - s_{i-1}}{s_i - s_{i-1}} F(w_i)) - \exp(\sqrt{-1}\pi \frac{t' - s_{i-1}}{s_i - s_{i-1}} F(w_i))\| \\ &\le \|\pi F(w_i)\| \frac{|t - t'|}{l_i/L} = L|t - t'|. \end{aligned} \tag{e 9.13}$$

One then computes that

$$\|v_t - v_{t'}\| \le L|t - t'| \text{ for all } t, t' \in [0,1]. \tag{e 9.14}$$

□

The following follows from Zhang's approximately halving projection lemma.

Lemma 9.4. *Let A be a non-elementary simple C^*-algebra of real rank zero and let $p \in A$ be a non-zero projection. Then, for any integer $n \ge 1$, there exist mutually orthogonal projections $p_1, p_2, ..., p_{n+1}$ such that p_1 is equivalent to p_i, $i = 1, 2, ..., n$, p_{n+1} is equivalent to a projection in $p_1 A p_1$ and*

$$p = \sum_{i=1}^{n+1} p_i.$$

Proof. Fix n. There is $m \ge 1$ such that $2^m \ge 2n^2$. Write

$$2^m = ns + r, \;\; r < s,$$

where $s \ge 1$ is a positive integer and r is a nonnegative integer. It follows that $s \ge n$.

By Theorem 1.1 of [**57**], there are mutually orthogonal projections $q_1, q_2, ...,$ q_{2^m+1} such that q_1 is equivalent to q_k, $k = 1, 2, ..., 2^m$, and q_{2^m+1} is equivalent to a projection in q_1Aq_1. Put

$$p_i = \sum_{k=(i-1)s+1}^{is} q_{k},, \quad i = 1, 2, ..., n \text{ and}$$

$$p_{n+1} = q_{2^m+1} + \sum_{k=ns+1}^{2ns+r} q_k.$$

One checks that so defined $p_1, p_2, ..., p_{n+1}$ meet the requirements. □

Theorem 9.5. *Let X be a connected finite CW complex with* $\dim(X) \geq 2$ *and let A be a unital separable simple C^*-algebra with real rank zero and stable rank one.*

Then there exists a sequence of unital monomorphisms $h_n : C(X) \to A$ and a sequence of unitaries $\{u_n\} \subset A$ satisfying the following:

(e 9.15) $$\lim_{n\to\infty} \|[u_n, h_n(a)]\| = 0 \text{ and } \mathrm{Bott}(h_n, u_n) = 0, n = 1, 2, ...,$$

for all $a \in C(X)$, and, for any $L > 0$, there exists ϵ_0 and a finite subset $\mathcal{F} \subset C(X)$ such that

(e 9.16) $$\sup\{\sup\{\|[u_n(t), h_n(a)]\| : t \in [0, 1]\} : a \in \mathcal{F}\} \geq \epsilon_0$$

for any sequence of continuous rectifiable path $\{u_n(t) : t \in [0, 1]\}$ of unitaries of A for which

(e 9.17) $$u_n(0) = u_n, \ u_n(1) = 1_A \text{ and } \mathrm{Length}(\{u_t\}) \leq L.$$

PROOF. For each integer n, there are mutually orthogonal projections e_n, $p(n, 1), p(n, 2), ..., p(n, n^3)$ in A such that $p(n, 1)$ is equivalent to every $p(n, i)$,$i =$ $2, 3, ..., n^3$, and

$$[e_n] \leq [p(n, 1)] \text{ and } e_n + \sum_{i=1}^{n^3} p(n, i) = 1_A.$$

by 9.4. Since $\dim X \geq 2$, there is a subset $X_0 \subset X$ which is homeomorphic to a (closed) k-cell with dimension of X_0 at least 2. There is a compact subset $Y \subset X_0$ such that Y is homeomorphic to S^1. Without loss of generality, we may write $Y = S^1$. By 2.25, we define a monomorphism $h_n^{(0)} : C([0, 1]) \to e_nAe_n$ and let $h_{00} : C(X) \to C([0, 1])$ be a monomorphism. Put $\psi_n = h_n^{(0)} \circ h_{00}$. Let $\{e_{i,j} : 1 \leq i, j \leq n^3\}$ be a system of matrix units for M_{n^3}. By mapping $e_{1,1}$ to $p(n, 1)$, we obtain a unital monomorphism $j_n : M_{n^3} \to (1 - e_n)A(1 - e_n)$. Put

$$E(n, i) = \sum_{j=n(i-1)+1}^{ni} p(n, j), \ \ i = 1, 2, ..., n^2.$$

j_n induces a unital homomorphism $j_{n,i} : M_n \to E(n, i)AE(n, i)$, $i = 1, 2, ..., n^2$. We will again use $j_{n,i}$ for the extension from $M_2(M_n)$ to $M_2(E(n, i)AE(n, i))$, $i = 1, 2, ..., n^2$. As in [**11**], [**12**] and [**43**] there are two sequences of unitaries $\{u_n'\}$ and $\{v_n'\}$ in M_n such that

(e 9.18) $$\lim_{n\to\infty} \|u_n'v_n' - v_n'u_n'\| = 0 \text{ and } \mathrm{rank}\{[e(u_n', v_n')]\} = n - 1.$$

Denote $p_n' = \begin{pmatrix} 1_{M_n} & 0 \\ 0 & 0 \end{pmatrix}$ and $q_n' = e(u_n', v_n')$ in $M_2(M_n)$. There are partial isometries $w_n' \in M_2(M_n)$ such that $p_n' - (w_n')^* q_n' w_n'$ is a rank one projection in $M_2(M_n)$.

Define $p_n = j_{n,i}(p_n')$, $q_n = j_{n,i}(q_n')$ and $w_n = j_{n,i}(w_n')$, $i = 1, 2, ..., n^2$ and $n = 1, 2, ...$ Define $\phi_{n,i}' : C(S^1) \to p(n,j)Ap(n,j)$ by $\phi_{n,i}'(g) = g(j_{n,i}(v_n'))$ for $g \in C(S^1)$, $n = 1, 2, ...,$ and define $\phi_{n,i} = \phi_{n,i}' \circ \pi'$, where $\pi' : C(X) \to C(S^1)$ is the quotient map which first maps $C(X)$ onto $C(X_0)$ and then maps $C(X_0)$ onto $C(S^1)$. Define $h_n : C(X) \to A$ by

$$h_n(f) = \psi_n(f) \oplus \sum_{i=1}^{n^2} \phi_{n,i}(f) \tag{e 9.19}$$

for $f \in C(X)$. Note that h_n are monomorphisms. Define

$$u_n = e_n \oplus \sum_{i=1}^{n^2} j_{n,i}(u_n'), \quad n = 1, 2, ... \tag{e 9.20}$$

Since each $j_{n,i}(u_n)$ is in a finite dimensional C^*-subalgebra,

$$[u_n] = 0 \text{ in } K_1(A). \tag{e 9.21}$$

We also have

$$\lim_{n\to\infty} \|[h_n(f), u_n]\| = 0 \text{ for all } f \in C(X). \tag{e 9.22}$$

Since ψ_n factors through $C([0,1])$ and $\phi_{n,i}$ factors through $C(X_0)$, by (e 9.21) and 9.2,

$$\mathrm{Bott}(h_n, u_n) = 0, \quad n = 1, 2, ... \tag{e 9.23}$$

Denote $B = l^\infty(A)$ and $Q = q_\infty(A)$. Let $H : C(X) \to B$ be defined by $\{h_n\}$. Let I_d be the (closed two-sided) ideal of B generated by

$$d = \{E(n,1)\}.$$

Put

$$\begin{aligned} p &= \{\mathrm{diag}(0, j_{n,1}(p_n'), j_{n,2}(p_n'), ..., j_{n,n^2}(p_n'))\}, \\ q &= \{\mathrm{diag}(0, j_{n,1}(q_n'), j_{n,2}(q_n'), ..., j_{n,n^2}(q_n'))\} \quad \text{and} \\ w &= \{\mathrm{diag}(0, j_{n,1}(w_n'), j_{n,2}(w_n'), ..., j_{n,n^2}(w_n'))\}. \end{aligned} \tag{e 9.24}$$

Put

$$e = p - w^* q w. \tag{e 9.25}$$

Note that $[E(n,1)] = n[p_n - w_n^* q_n w_n]$. Hence

$$[e] = n[E(n,1)].$$

It follows from 9.1 that

$$e \notin M_2(I_d) \tag{e 9.26}$$

Now suppose that the lemma fails. Then, by passing to a subsequence if necessary, we obtain a constant $L > 0$ and a sequence of continuous rectifiable paths $\{z_t^{(n)} : t \in [0,1]\}$ of A such that

$$z_0^{(n)} = u_n, \quad z_1^{(n)} = 1_A \text{ and } \lim_{n\to\infty} \|[z_t^{(n)}, h_n(f)]\| = 0. \tag{e 9.27}$$

Moreover, for each n,

$$\text{Length}(\{z_t^{(n)}\}) \le L \tag{e 9.28}$$

It follows from 9.3 that there exists, for each n, a path of unitaries $\{v_t(n) : t \in [0,1]\}$ in A such that

$$v_0(n) = z_0^{(n)}, \ \lim_{n\to\infty} \|[h_n(g), v_t(n)]\| = 0$$

for all $g \in C(X)$ and

$$\|v_t(n) - v_{t'}(n)\| \le L|t - t'| \tag{e 9.29}$$

for all $t, t' \in [0,1]$ and all n. In particular, $\{v_t(n)\}$ is equi-continuous on $[0,1]$.

Let $\pi : B \to Q$ be the quotient map. Put $Z_t = \pi(\{v_t\})$ for $t \in [0,1]$. By (e 9.29), $\{Z_t : t \in [0,1]\}$ is a continuous path of unitaries in Q. Moreover

$$Z_0 = \pi(\{u_n\}), \ \ Z_1 = 1_B \ \text{ and } \ Z_t(\pi \circ H(f)) = (\pi \circ H(f))Z_t \tag{e 9.30}$$

for all $f \in C(X)$ and $t \in [0,1]$. Let $\bar{\pi}_I : Q \to Q/\pi(I_d)$. Then $\bar{\pi}_I \circ \pi \circ H$ induces a homomorphism $\Psi : C(S^1) \to Q/\pi(I_d)$ since $\{e_n\} \in I_d$. We also have

$$\bar{\pi}_I(Z_0) = \bar{\pi}_I(\{u_n\}), \ \ \bar{\pi}_I(Z_1) = 1 \ \text{ and } \ \bar{\pi}_I(Z_t)\Psi(g) = \Psi(g)\bar{\pi}_I(Z_t\}) \tag{e 9.31}$$

for all $g \in C(S^1)$. However, since $e \notin M_2(I_d)$, by (e 9.18), we compute that

$$\text{bott}_1(\Psi, \bar{\pi}_I(\{u_n\}) = [\bar{\pi}_I(e)] \ne 0. \tag{e 9.32}$$

On the other hand, from (e 9.31),

$$\text{Bott}(\Psi, \bar{\pi}_I(\{u_n\})) = \text{Bott}(\Psi, \bar{\pi}_I(Z_t)) = \text{Bott}(\Psi, \bar{\pi}_I(Z_1)) = 0.$$

This contradicts (e 9.32). □

9.6. As we know that a continuous rectifiable path with length L may fail to be Lipschitz, for the future use, applying 9.3, Theorem 8.3 may be written as the following. We also remark that Theorem 3.13 may also be written similarly.

Theorem 9.7. *Let X be a compact metric space. For any $\epsilon > 0$, any finite subset $\mathcal{F} \subset C(X)$ and any (increasing) map $\Delta : (0,1) \to (0,1)$, there exists $\delta > 0$, a finite subset $\mathcal{G} \subset C(X)$ and a finite subset $\mathcal{P} \subset \underline{K}(C(X))$ satisfying the following:*

Suppose that A is a unital separable simple C^*-algebra with tracial rank zero, $h : C(X) \to A$ is a unital monomorphism and $u \in A$ is a unitary such that $\mu_{\tau\circ h}$ is Δ-distributed for all $\tau \in T(A)$,

$$\|[h(a), u]\| < \delta \ \textit{for all} \ a \in \mathcal{G} \ \textit{and} \ \text{Bott}(h, u)|_{\mathcal{P}} = 0. \tag{e 9.33}$$

Then, there exists a continuous rectifiable path of unitaries $\{u_t : t \in [0,1]\}$ of A such that

$$u_0 = u, \ \ u_1 = 1_A \ \textit{and} \ \|[h(a), u_t]\| < \epsilon \ \text{for all} \ a \in \mathcal{F} \ \textit{and} \ t \in [0,1]. \tag{e 9.34}$$

Moreover,

$$\|u_t - u_{t'}\| \le (2\pi + \epsilon)|t - t'| \ \textit{for all} \ t, t' \in [0,1] \tag{e 9.35}$$

$$\textit{and} \ \text{Length}(\{u_t\}) \le 2\pi + \epsilon. \tag{e 9.36}$$

Remark 9.8. Let X be a connected finite CW complex, let A be a unital simple C^*-algebra and let $h : C(X) \to A$ be a unital monomorphism. Suppose that $u \in A$ is a unitary such that $[h(f), u] = 0$ for all $f \in C(X)$. Denote by $\phi : C(X \times S^1) \to A$ the homomorphism induced by h and u. Suppose there is a rectifiable continuous path of unitaries $\{u_t : t \in [0, 1]\}$ of A such that $u_0 = u$, $u_1 = 1$ and $\|[h(a), u]\| < \epsilon < 1/2$ on a set of generators. We know that it is necessary that $\mathrm{Bott}(h, u) = 0$. Theorem 6.1 states that under the assumption that ϕ is injective, $\mathrm{Bott}(h, u) = 0$ is also sufficient to have such path of unitaries provided that A has tracial rank zero.

As in the proof of 9.5, if there is a sequence of unitaries $u_n \in A$ such that

$$\lim_{n\to\infty} \|[h(a), u_n]\| = 0 \text{ for all } f \in C(X),$$

then we obtain a homomorphism $\Psi : C(X \times S^1) \to q_\infty(A)$. The existence of the desired paths of unitaries implies (at least) that

$$\Psi_{*0}|_{\beta_1(K_1(C(X)))} = 0.$$

However, $q_\infty(A)$ is not simple. Suppose that $I \subset q_\infty(A)$ is a closed two-sided ideal and $\bar{\pi} : q_\infty(A) \to q_\infty(A)/I$ is the quotient map. Even if Ψ is a monomorphism, $\mathrm{ker}\bar{\pi} \circ \Psi$ may not be zero. The trouble starts here. Suppose that F is a compact subset of X such that

$$\mathrm{ker}\Psi \supset \{f \in C(X \times S^1) : f|_{F\times S^1} = 0\}.$$

We obtain a homomorphism $\Psi_1 : C(F \times S^1) \to q_\infty(A)/I$. The existence of the paths also requires that

$$(\Psi_1)_{*0}(\beta_1(K_1(C(F)))) = 0.$$

This can not be guaranteed by $\Psi_{*0}|_{\beta_1(K_1(C(X)))} = 0$ when $\mathrm{dim}X \geq 2$ as we see in the proof. The relevant ideals are those associated with diminishing tracial states. This topological obstruction can be revealed this way. The requirement to have a fixed measure distribution as in 7.4 guarantees that $\mathrm{ker}\bar{\pi} \circ \Psi = \{0\}$ for those ideals. Therefore the condition of vanishing Bott maps is sufficient at least for the case that A is assumed to have tracial rank zero.

When $\mathrm{dim}X \leq 1$, $s_{*1}(K_1(C(X))) = K_1(C(F))$, where $s : C(X) \to C(F)$ is the quotient map. Thus ideals with diminishing tracial states do not produce anything surprising. Therefore, when $\mathrm{dim}X \leq 1$, as one sees in the section 3, no measure distributions are considered.

In Section 11, we will consider the case that A is assumed to be purely infinite and simple. In that case, A has no tracial states. So there is nothing about measure distribution. In turns out, there is no hidden obstruction as we will see in the section after the next.

CHAPTER 3

Purely infinite simple C^*-algebras

10. Purely infinite simple C^*-alegbras

Recall that a unital C^*-algebra A is said to be purely infinite and simple, if $A \neq \mathbb{C}$ and, for any nonzero element $a \in A$, there are $x, y \in A$ such that $xay = 1_A$. (see [**5**]). Well-known examples of purely infinite simple C^*-algebras include the Cuntz algebras $\mathcal{O}_n$, $\mathcal{O}_\infty$ and the Calkin algebra. Amenable separable purely infinite simple C^*-algebras which satisfy the UCT become one of the best understood classes of C^*-algebras thanks to Kirchberg and Phillips's classification theorem (see [**47**] and [**50**], for example).

Lemma 10.1 (Lemma 1.5 of [**40**]). *Let X be a compact metric space, let $\mathcal{F}$ be a finite subset of the unit ball of $C(X)$. For any $\epsilon > 0$ and any $\sigma > 0$, there exists $\eta > 0$ such that for any unital C^*-algebra A of real rank zero and any contractive unital $*$-preserving linear map $\psi : C(X) \to A$, there is a finite subset $\{x_1, x_2, ..., x_n\} \subset \Sigma_\eta(\psi, \mathcal{F})$ which is σ-dense in $\Sigma_\eta(\psi, \mathcal{F})$ (see 2.22) and mutually orthogonal nonzero projections $p_1, p_2, ..., p_n$ in A satisfying the following:*

$$\|\psi(f) - \sum_{i=1}^{n} f(x_i)p_i - p\psi(f)p\| < \epsilon$$

for all $f \in \mathcal{F}$, where $p = 1 - \sum_{i=1}^n p_i$.

PROOF. The statement is slightly different from that of Lemma 1.5 of [**40**] and can be deduced from it. □

Lemma 10.2. *Let X be a compact metric space and let $\mathcal{F} \subset C(X)$ be a finite subset. For any $\epsilon > 0$ and any $\sigma > 0$, there exists $\delta > 0$ and a finite subset $\mathcal{G} \subset C(X)$ satisfying the following:*

For any unital C^*-algebra B and any unital monomorphism $\phi : C(X) \to B$, if there exists a unitary $u \in B$ such that

$$\|[\phi(g), u]\| < \delta \text{ for all } g \in \mathcal{G},$$

then there is an ϵ-$\mathcal{F} \otimes S$-multiplicative contractive completely positive linear map $\psi : C(X \times S^1) \to B$ for which

$$\|\psi(f \otimes g) - \phi(f)g(u)\| < \epsilon$$

for all $f \in \mathcal{F}$ and $g \in S$ and for which there is a finite σ-dense subset $\{x_1, x_2, ..., x_m\}$ $\subset X$ and a finite subset $\{s_1, s_2, ..., s_m\} \subset S^1$ (may be not distinct) such that $(x_i, s_i) \in \Sigma_\epsilon(\psi, \mathcal{F} \otimes S)$, $i = 1, 2, ..., m$, where $S = \{1_{C(S^1)}, z\}$.

PROOF. The proof is a combination of 2.8 and the proof of 2.23. Without loss of generality, we may assume that $\mathcal{F}$ is a finite subset of the unit ball of $C(X)$. Fix a σ-dense subset $\{x_1, x_2, ..., x_m\}$ in X. Let $\eta = \sigma_{X,\epsilon/3,\mathcal{F}}$. In particular

$$|f(x) - f(x')| < \epsilon/3 \text{ for all } f \in \mathcal{F},$$

whenever $\mathrm{dist}(x, x') < \eta$. Choose non-negative functions $f_1, f_2, ..., f_m, f_1', f_2', ..., f_m' \in C(X)$ such that $f_i(x) \le 1$, $f_i(x) = 1$ if $x \in O(x_i, \eta/4)$ and $f_i(x) = 0$ if $x \notin O(x_i, \eta/2)$ and $f_i'(x) \le 1$, $f_i'(x) = 1$ if $x \in O(x_i, \eta/8)$ and $f_i'(x) = 0$ if $x \notin O(x_i, \eta/4)$, $i = 1, 2, ..., m$. Choose $n \ge 1$ such that $2\pi/n < \epsilon/24$. Let $I_j' \subset S^1$ be an open arc with center at $2j\pi/n$ and with length $4\pi/n$, $j = 1, 2, ..., n$. Note that $\cup_{j=1}^n I_j' = S^1$. Let $\{g_j' : 1 \le j \le n\} \subset C(S^1)$ be a partition of unit associated with $\{I_1', I_2', ..., I_n'\}$. Let I_j be an open arc with center at $2j\pi/n$ and with length $8\pi/n$, $j = 1, 2, ...n$. Let $g_j \in C(S^1)$ such that $0 \le g_j(t) \le 1$, $g_j(t) = 1$ if $t \in I_j'$ and $g_j(t) = 0$ if $t \notin I_j$, $j = 1, 2, ..., n$.

There is $\delta_1 > 0$ and a finite subset $\mathcal{G}_1 \subset C(X)$ satisfying the following: if $\phi' : C(X) \to B$ is a unital homomorphism and $u' \in B$ is a unitary (for any unital C^*-algebra B) for which

$$\|[\phi'(f), u']\| < \delta_1 \text{ for all } f \in \mathcal{G}_1,$$

then

$$\|\phi'(f_i')g_j'(u') - g_j'(u')\phi'(f_i')\| < \epsilon/24, \ \ i = 1, 2, ..., m \text{ and } j = 1, 2, ..., n.$$

Choose $\delta > 0$ and a finite subset $\mathcal{G} \subset C(X)$ as required by 2.8 for $\min\{\delta_1, \eta, \epsilon/24\}$ and $\mathcal{F} \cup \mathcal{G}_1$. We may also assume that $\mathcal{G} \supset \mathcal{F} \cup \mathcal{G}_1$. Put $\epsilon_1 = \min\{\delta_1, \eta, \epsilon/24\}$. It follows from 2.8 that there exists an ϵ_1-$(\mathcal{F} \cup \mathcal{G}_1) \otimes S$-multiplicative contractive completely positive linear map $\psi : C(X \times S^1)$ such that

$$\|\psi(f \otimes g) - h(f)g(u)\| < \epsilon_1 \text{ for all } f \in \mathcal{F} \cup \mathcal{G}_1 \text{ and } \text{ for all } g \in S.$$

Put $t_j = 2j\pi/n$, $j = 1, 2, ..., n$.

Now suppose that ϕ and u satisfy the assumptions of the lemma with the above δ and $\mathcal{G}$. For each i, since $\sum_{j=1}^n g_j'(u) = 1$, there exists j such that $\phi(f_i')g_j'(u) \ne 0$. Let $B_i = \overline{\phi(f_i')g_j'(u)Bg_j'(u)\phi(f_i')}$, $i = 1, 2, ..., m$. Let $b = \phi(f_i')g_j'(u)cg_j'(u)\phi(f_i')$ for some $c \in B$ with $\|b\| \le 1$. Then, by the choices of ϵ_1 and $\mathcal{G}_1$, we estimate that

$$\begin{aligned}
&\|(f(x_i)g(t_j) - \psi(f \otimes g))b\| \\
&\quad \le \ \epsilon_1 + \|(f(x_i)g(t_j) - \phi(f)g(t_j))b\| + \|(\phi(f)g(t_j) - \phi(f)g(u))b\| \\
&\quad < \ \epsilon/24 + \|g(t_j)[f(x_i) - \phi(f)]\phi(f_i)b\| + \epsilon/24 + \\
&\qquad\qquad \|\phi(f)\|\|(g(t_j) - g(u))g_j'(u)\phi(f_i')cg_j'(u)\phi(f_i')\| \\
&\quad < \ \epsilon/12 + \epsilon/3 + \|(g(t_j) - g(u))g_j(u)g_j'(u)\phi(f_i')cg_j'(u)\phi(f_i')\| \\
&\quad < \ \epsilon/12 + \epsilon/3 + \epsilon/24 + \|(g(t_j) - g(u))g_j(u)b\| \\
&\quad < \ \epsilon/12 + \epsilon/3 + \epsilon/24 + \epsilon/6 < \frac{15\epsilon}{24}.
\end{aligned}$$

It follows that

$$\|(f(x_i)g(t_j) - \psi(f \otimes g))b\| < \epsilon$$

for all $b \in B_i$ with $\|b\| \le 1$, for all $f \in \mathcal{F}$ and $g \in S$. The lemma follows. □

Theorem 10.3. *Let X be a compact metric space and let $\epsilon > 0$ and let $\mathcal{F} \subset C(X)$ be a finite subset. There is $\delta > 0$ and a finite subset $\mathcal{G} \subset C(X)$ and a finite subset $\mathcal{P} \subset \underline{K}(C(X))$ satisfying the following:*

Suppose that A is a unital purely infinite simple C^*-algebra and $\psi, \phi : C(X) \to A$ are two unital δ-$\mathcal{G}$-multiplicative contractive completely positive linear maps for which

$$[\psi]|_{\mathcal{P}} = [\phi]|_{\mathcal{P}}, \tag{e 10.1}$$

and ψ and ϕ are both $1/2$-$\mathcal{G}$-injective, then exists a unitary $u \in A$ such that

$$\|\mathrm{ad}\, u \circ \psi(f) - \phi(f)\| < \epsilon \text{ for all } f \in \mathcal{F}. \tag{e 10.2}$$

PROOF. If A is also assumed to be amenable, this theorem follows from the combination of a theorem of Dadarlat (Theorem 1.7 of [**6**]) and Cor 7.6 of [**32**]. In the general case, we proceed as follows.

Let $\delta_1 > 0$ and $\mathcal{G}_1$ and $\mathcal{P}$ be required as in Theorem 3.1 of [**14**] for $\epsilon/16$ and $\mathcal{F}$. Note (as stated before Theorem 3.1 of [**14**]) that Theorem 3.1 of [**14**] applies to the case that A is a unital purely infinite simple C^*-algebra. We may assume that $\mathcal{F} \subset \mathcal{G}_1$.

By choosing smaller δ_1 and larger $\mathcal{G}_1$, we may assume that

$$[L_1]|_{\mathcal{P}} = [L_2]|_{\mathcal{P}}$$

if both L_1 and L_2 are δ_1-$\mathcal{G}_1$-multiplicative contractive completely positive linear map from $C(X)$ to A, and if

$$L_1 \approx_{\delta_1} L_2 \text{ on } \mathcal{G}_1.$$

Let $\eta_1 = \min\{\epsilon/16, \delta_1/4\}$ and $\eta_2 < \sigma_{X,\mathcal{G}_1,\eta_1}$. Therefore $|f(x) - f(y)| < \eta_1$ for all $f \in \mathcal{G}_1$, whenever $\mathrm{dist}(x, y) < \eta_2$. Put $\eta_3 = \min\{\eta_1, \eta_2\}$. Let $\eta > 0$ be required in 10.1 corresponding to η_3 (in place of ϵ), $\eta_2/2$ (in place of σ) and $\mathcal{G}_1$ (in place of $\mathcal{F}$). We may assume that $\eta < \eta_3/2$.

Let $\delta > 0$ and $\mathcal{G} \subset C(X)$ be a finite subset required by 2.23 associated with η (in place of ϵ) and $\mathcal{G}_1$ (in place of $\mathcal{F}$). We may assume that $\mathcal{G}_1 \subset \mathcal{G}$ and $\delta < \eta$.

Now suppose that $\psi, \phi : C(X) \to A$ are two unital δ-$\mathcal{G}$-multiplicative contractive completely positive linear maps which are both $1/2$-$\mathcal{G}$-injective such that

$$[\psi]|_{\mathcal{P}} = [\phi]|_{\mathcal{P}}. \tag{e 10.3}$$

By Lemma 1.5 of [**40**] and 2.23,

$$\|\phi(f) - \sum_{j=1}^{N} f(\xi_j)p_j + \phi_1(f)\| < \eta_3 \text{ and } \|\psi(f) - \sum_{i=1}^{N'} f(\xi_i')p_i' + \psi_1(f)\| < \eta_3 \tag{e 10.4}$$

for all $f \in \mathcal{G}_1$, where $\{p_1, p_2, ..., p_N\}$ and $\{p_1, p_2, ..., p_{N'}\}$ are two sets of non-zero mutually orthogonal projections, $P = \sum_{j=1}^{N} p_j$, $P' = \sum_{i=1}^{N'} p_i'$, $\{\xi_1, \xi_2, ..., \xi_N\}$ and $\{\xi_1', \xi_2', ..., \xi_{N'}'\}$ are two η_2-dense subsets in X and $\psi_1 : C(X) \to (1 - P)A(1 - P)$ and $\psi_1' : C(X) \to (1 - P')A(1 - P')$ are two unital δ-$\mathcal{G}$-multiplicative contractive completely positive linear maps.

By the choice of η_2 and by replacing η_3 by η_1 in (e 10.4), we may assume that $N = N'$ and $\xi_i = \xi_i'$, $i = 1, 2, ..., N$. Since A is purely infinite, we may write $p_i = e_i + e_i'$, where e_i, e_i' are two mutually orthogonal projections such that $[e_i] = [p_i']$, $i = 1, 2, ..., N$. Put $\phi_1'(f) = \sum_{i=1}^{N} f(\xi_i)e_i' + \phi_1(f)$ for all $f \in C(X)$. There is a unitary $U \in A$ such that

$$U^* p_i' U = e_i, \ \ i = 1, 2, ..., N.$$

Replacing ϕ_1 by ϕ_1' and ψ by $\mathrm{ad}\, U \circ \psi$, in (e 10.4), we may assume that $p_i = p_i'$ and $P = P'$. In other words, we may assume that

$$\text{(e 10.5)} \quad \|\phi(f) - \sum_{j=1}^{N} f(\xi_j)p_j + \phi_1(f)\| < \eta_1 \text{ and } \|\psi(f) - \sum_{i=1}^{N} f(\xi_i)p_i + \psi_1(f)\| < \eta_1$$

for all $f \in \mathcal{G}_1$,

Define $h_0 : C(X) \to PAP$ by $h_0(f) = \sum_{i=1}^N f(\xi_i)p_i$ for all $f \in C(X)$. By the choice of δ_1 and $\mathcal{G}_1$ and by (e 10.3), ϕ_1 and ψ_1 are $\delta_1/2$-$\mathcal{G}_1$-multiplicative and

$$\text{(e 10.6)} \qquad [\phi_1]|_{\mathcal{P}} = [\psi_1]|_{\mathcal{P}}.$$

By Theorem 3.1 of [**14**], there is an integer $l > 0$, a unitary $w \in M_{l+1}((1-P)A(1-P))$ and a unital homomorphism $h_{00} : C(X) \to M_l((1-P)A(1-P))$ with finite dimensional range such that

$$\text{(e 10.7)} \qquad \mathrm{diag}(\psi_1, h_{00}) \approx_{\epsilon/16} \mathrm{ad}\, w \circ \mathrm{diag}(\phi_1, h_{00}) \text{ on } \mathcal{F}.$$

By the choice of η_2 and by replacing $\epsilon/16$ by $\epsilon/8$ in (e 10.7), we may assume that

$$\text{(e 10.8)} \qquad h_{00}(f) = \sum_{j=1}^{N} f(\xi_j)q_j \text{ for all } f \in C(X)$$

where $\{q_1, q_2, ..., q_N\}$ are mutually orthogonal projections in $M_l((1-P)A(1-P))$. Since A is purely infinite and simple, we may write that $p_i = q_i' + q_i''$, where q_i and q_i' are two mutually orthogonal projections such that $[q_i] = [q_i']$, $1 = 2, ..., N$. There is a unitary $w_1 \in M_{l+1}(A)$ such that

$$\text{(e 10.9)} \qquad w_1^* q_i w = q_i' \le p_i, \quad i = 1, 2, ..., N.$$

Thus, by (e 10.7), we obtain a unitary $u \in A$ such that

$$\text{(e 10.10)} \qquad \mathrm{ad}\, u \circ (h_0 + \phi_1) \approx_{\epsilon/4} h_0 + \psi_1 \text{ on } \mathcal{F}.$$

By (e 10.5), we have

$$\mathrm{ad}\, u \circ \psi \approx_\epsilon \phi \text{ on } \mathcal{F}.$$

□

Corollary 10.4. *Let X be a compact metric space and let $\mathcal{F} \subset C(X)$ be a finite subset. For any $\epsilon > 0$, there are $\delta > 0$, a finite subset $\mathcal{G} \subset C(X)$ and a finite subset $\mathcal{P} \subset \underline{K}(C(X))$ satisfying the following:*

For any unital purely infinite simple C^*-algebra A and any unital δ-$\mathcal{G}$-multiplicative contractive completely positive linear map $\psi : C(X) \to A$ which is $1/2$-$\mathcal{G}$-injective such that

$$[\psi]|_{\mathcal{P}} = [h]|_{\mathcal{P}},$$

for some homomorphism $h : C(X) \to A$, there is a unital monomorphism $\phi : C(X) \to A$ such that

$$\|\psi(f) - \phi(f)\| < \epsilon \text{ for all } f \in \mathcal{F}.$$

Proof. For sufficiently small δ and large $\mathcal{G}$, we may assume that $[h(1_{C(X)})] = [1_A]$. To make h injective, let $e \in A$ be a projection with $[e] = [h(1_{C(X)})]$ such that $1 - e \neq 0$. So we may assume that h maps $C(X)$ unitally to eAe. But then $[1-e] = 0$. There is a unital embedding $\phi_0 : \mathcal{O}_2 \to eAe$. By applying 2.25, we obtain a unital embedding $h_0 : C(X) \to \mathcal{O}_2$. Put $h_{00} = \phi_0 \circ h_0$. Then $h_1 = h + h_{00}$. Then h_1 is a unital monomorphism. We then apply 10.3 to obtain ϕ. $\square$

The following is a combination of a result of M. Rørdam [**49**], a result of [**10**] and the classification theorem of Kirchberg-Phillips .

Lemma 10.5. *Let A be a unital separable amenable purely infinite simple C^*-algebra satisfying the Universal Coefficient Theorem and B be any unital purely infinite simple C^*-algebra. Then for any $\alpha \in KL(A, B)$, there is a monomorphism $\psi : A \to B$ such that*

$$\alpha = [\psi].$$

Moreover, if $\alpha([1_A]) = [1_B]$, ψ can be made into a unital homomorphism.

PROOF. Let $\bar{\mathcal{C}}$ be the class of those unital separable purely infinite simple C^*-algebras for which the lemma holds. Let (G_0', g_0', G_1') be a triple, where G_0' and G_1' are finitely generated abelian groups and $g_0' \in G_0'$ is an element. The proof of 5.6 of [**10**] shows that the class $\mathcal{C}$ contains one unital separable amenable purely infinite simple C^*-algebra A satisfying the UCT such that $(K_0(A), [1_A], K_1(A)) = (G_0', g_0', G_1')$. Now let (G_0, g_0, G_1) be a triple for arbitrary countable abelian groups. Then it is an inductive limit of those triples with finitely generated abelian groups. By 8.2 of [**49**] (see also 5.9 of [**49**] and 5.4 of [**10**]), $\bar{C}$ contains a unital separable amenable purely infinite simple C^*-algebra A satisfying the UCT such that $(K_0(A), [1_A], K_1(A)) = (G_0, g_0, G_1)$. By the classification of those purely infinite simple C^*-algebras of Kirchberg and Phillips (see [**47**], for example), there is only one such unital amenable separable purely infinite simple C^*-algebra satisfying the UCT. $\square$

The following also follows from a result of E. Kirchberg (in a coming paper).

Theorem 10.6. *Let B be a unital separable amenable C^*-algebra which satisfies the UCT and let A be a unital purely infinite simple C^*-algebra. Then, for any $\kappa \in KL(B, A)$, there is a monomorphism $h : B \to A$ such that*

$$[h] = \kappa.$$

If A is unital and $\kappa([1_B]) = [1_A]$, then h can be chosen to be unital.

PROOF. There is a unital amenable separable purely infinite simple C^*-algebra C satisfying the UCT such that

$$(K_0(C), [1_C], K_1(C)) = (K_0(B), [1_B], K_1(B)).$$

Thus, by 7.2 of [**51**], there is an invertible element $x \in KK(B, C)$. Thus, by Theorem 6.7 of [**32**], there is a unital homomorphism $h_1 : B \to C$ such that $[h_1] = \bar{x}$, where $\bar{x} \in KL(B, C)$ is the image of x. Since x is invertible, we obtain an isomorphism

$$KK(B, A) \cong KK(C, A).$$

Let $\tilde{\kappa} \in KK(B, A)$ such that the image of $\tilde{\kappa}$ in $KL(B, A)$ is κ. Let $\tilde{\kappa}' \in KK(C, A)$ be the element which is the image of $\tilde{\kappa}$ under the above isomorphism. It follows that there is $\tilde{\alpha} \in KK(C, C)$ such that

$$[h_1] \times \tilde{\alpha} \times \tilde{\kappa}' = \tilde{\kappa}, \tag{e 10.11}$$

where $\times$ is the Kasparov product. Let α be the image of $\tilde{\alpha}$ in $KL(C,C)$ and let κ' be the image of $\tilde{\kappa}'$ in $KL(C,A)$, respectively. By 10.5, there are a monomorphism $h_2 : C \to A$ and a homomorphism $h_3 : C \to C$ such that

$$[h_2] = \kappa' \text{ and } [h_3] = \alpha. \tag{e 10.12}$$

Define $h' = h_2 \circ h_3 \circ h_1$. Then

$$[h'] = \kappa \text{ in } KL(B, A).$$

To make sure that h' is a monomorphism, consider two non-zero projections $p_1, p_2 \in A$ such that $p_1 \le p_2$, $p_2 - p_1 \ne 0$ and

$$[p_1] = [p_2] = \kappa([1_B]).$$

By applying what we have shown to $p_1 A p_1$, we obtain a homomorphism $h' : B \to p_1 A p_1$ such that $[h'] = \kappa$.

There is a unital embedding h_0 from $\mathcal{O}_2$ into $(p_2 - p_1)A(p_2 - p_1)$ (note that $[p_2 - p_1] = 0$ in $K_0(A)$). It follows from 2.3 of [**17**] that there is a unital monomorphism $h_0' : B \to \mathcal{O}_2$. Let $h_{00} = h_0 \circ h_0'$. Then

$$[h_{00}] = 0 \text{ in } KK(B, A).$$

Define $h = h_{00} + h'$. Then h is a monomorphism. Moreover, if $\kappa([1_B]) = [1_A]$, we can choose $p_2 = 1_A$. □

Theorem 10.7. *Let X be a finite CW complex, let $\epsilon > 0$ and let $\mathcal{F} \subset C(X)$ be a finite subset. There exists $\delta > 0$ and a finite subset $\mathcal{G} \subset C(X)$ satisfying the following.*

For any purely infinite simple C^*-algebra A and any unital δ-$\mathcal{G}$ -multiplicative contractive completely positive linear map $\psi : C(X) \to A$ which is $1/2$-$\mathcal{G}$-injective, there is a unital monomorphism $h : C(X) \to A$ such that

$$\|\psi(f) - h(f)\| < \epsilon \text{ for all } f \in \mathcal{F}. \tag{e 10.13}$$

Proof. If we also assume that $A \otimes \mathcal{O}_\infty \cong A$, for example, A is amenable, this follows from Cor 7.6 of [**32**]. The general case follows from 10.4. Since $K_i(C(X))$ is finitely generated ($i = 0, 1$), it is easy to see that, with sufficiently small δ and sufficiently large $\mathcal{G}$, $[\psi]$ gives an element in $KL(C(X), A)$. This certainly follows from 2.4. The next thing is to show that, for any $\kappa \in KL(C(X), A)$, there is a unital monomorphism $h : C(X) \to A$ such that $[h] = \kappa$. This follows from Theorem 10.6. Then 10.4 applies. □

Lemma 10.8. *Let X be a compact metric space, let $\epsilon > 0$, let $\sigma > 0$, and let $\mathcal{F} \subset C(X)$ be a finite subset. Suppose that A is a unital purely infinite simple C^*-algebra and suppose that $h : C(X) \to A$ is a unital homomorphism with the spectrum $F \subset X$. Then there are nonzero mutually orthogonal projections $e_1, e_2, ..., e_n$ and a σ-dense subset $\{x_1, x_2, ..., x_n\} \subset F$ and a unital homomorphism $h_1 : C(X) \to (1-e)A(1-e)$ (where $e = \sum_{i=1}^n e_i$) with the spectrum F such that*

$$\|h(f) - (h_1(f) + \sum_{i=1}^n f(x_i)e_i)\| < \epsilon \text{ for all } f \in \mathcal{F}.$$

Moreover, we may choose $[e_i] = 0$ in $K_0(A)$, $i = 1, 2, .., n$ and $[h_1] = [h]$ in $KL(C(X), A)$.

PROOF. To simplify the notation, without loss of generality, we may assume that $X = F$. Fix $\epsilon > 0$ and a finite subset $\mathcal{F} \subset C(X)$. Without loss of generality, we may assume that $\mathcal{F}$ is in the unit ball of $C(X)$. Let $\delta > 0$ and $\mathcal{G} \subset C(X)$ be a finite subset required by 10.4 associated with $\epsilon/4$ and $\mathcal{F}$. Let $\delta_1 = \min\{\delta, \epsilon/4\}$. Let $\eta = \sigma_{X,\mathcal{G},\delta_1}$.

By choosing smaller δ and larger $\mathcal{G}$, we may also assume that

$$[L_1]|_{\mathcal{P}} = [L_2]|_{\mathcal{P}}$$

for any two unital δ-$\mathcal{G}$-multiplicative contractive completely positive linear maps from $C(X)$ to A, provided that

$$L_1 \approx_\delta L_2 \text{ on } \mathcal{G}.$$

Let $\{x_1, x_2, ..., x_n\}$ be a σ-dense subset of $X = F$ such that

$$\max\{|g(x_i)| : i = 1, 2, ..., n\} \geq (3/4)\|g\| \text{ for all } g \in \mathcal{G}. \tag{e 10.14}$$

It follows from 10.1 that there are non-zero mutually orthogonal projections $e_1, e_2, ..., e_n$ such that

$$\|h(f) - [(1-p)h(f)(1-p) + \sum_{i=1}^{n} f(x_i)e_i]\| < \delta_1 \text{ and} \tag{e 10.15}$$

$$\|[1-p, h(f)]\| < \delta_1 \tag{e 10.16}$$

for all $f \in \mathcal{G}$. Put $e = \sum_{i=1}^n e_i$. Define $\psi : C(X) \to (1-e)A(1-e)$ by $\psi(f) = (1-p)h(f)(1-p)$ for $f \in C(X)$. Then ψ is δ_1-$\mathcal{G}$-multiplicative, by (e 10.16). Write $e_i = e_i' + e_i''$, where e_i' and e_i'' are nonzero mutually orthogonal projections such that

$$[e_i'] = 0 \text{ in } K_0(A),\ \ i = 1, 2, ..., n. \tag{e 10.17}$$

This is possible since A is purely infinite and simple. By replacing $(1-p)h(f)(1-p)$ by $(1-p)h(f)(1-p) + \sum_{i=1}^n f(x_i)e_i''$ and replacing e_i by e_i', we may assume, by (e 10.14), that ψ is 1/2-$\mathcal{G}$-injective and $[e_i] = 0$ in $K_0(A)$, $i = 1, 2, ..., n$. Define $h_0 : C(X) \to eAe$ by $h_0(f) = \sum_{i=1}^n f(x_i)e_i$ for all $f \in C(X)$. Since $[e_i] = 0$ in $K_0(A)$, we have that

$$[h_0] = 0 \text{ in } KL(C(X), A). \tag{e 10.18}$$

Thus, by (e 10.15), (e 10.17) and by (e 10.18),

$$[\psi]|_{\mathcal{P}} = [h]|_{\mathcal{P}}. \tag{e 10.19}$$

It follows from 10.4 that there is a unital monomorphism $h_1 : C(X) \to (1-e)A(1-e)$ such that

$$\|h_1(f) - \psi(f)\| < \epsilon/4 \text{ for all } f \in \mathcal{F}. \tag{e 10.20}$$

Combining (e 10.15) and (e 10.20), we obtain that

$$\|h(f) - [h_1(f) + \sum_{i=1}^{n} f(x_i)e_i]\| < \epsilon \text{ for all } f \in \mathcal{F}.$$

This proves the first part of the lemma. Note that h_1 was chosen to have $[h_1] = [h]$ as in the proof of 10.4. Thus the above proof also implies the last part of the lemma. □

Corollary 10.9. *Let X be a compact metric space, let $\epsilon > 0$, let $\sigma > 0$, and let $\mathcal{F} \subset C(X)$ be a finite subset. Suppose that A is a unital purely infinite simple C^*-algebra and suppose that $h : C(X) \to A$ is a unital homomorphism with the spectrum $F \subset X$. Then, for any finite subset $\{x_1, x_2, ..., x_n\} \subset F$, any $z_1, z_2, ..., z_n \in K_0(A)$, there are nonzero mutually orthogonal projections $e_1, e_2, ..., e_n$ with $[e_i] = z_i$, $i = 1, 2, ..., n$, and a unital homomorphism $h_1 : C(X) \to (1-e)A(1-e)$ (where $e = \sum_{i=1}^n e_i$) with the spectrum F such that*

$$\|h(f) - (h_1(f) + \sum_{i=1}^n f(x_i)e_i)\| < \epsilon \text{ for all } f \in \mathcal{F}.$$

PROOF. The only thing that needs to be added to 10.8 is that there is a non-zero projection $e_i' \le e_i$ (in A) such that $[e_i'] = z_i$, $i = 1, 2, ..., n$. □

11. Basic Homotopy Lemma in purely infinite simple C^*-algebras

Now we establish the Basic Homotopy Lemma in purely infinite simple C^*-algebras.

Lemma 11.1. *Let X be a compact metric space, let A be a unital purely infinite simple C^*-algebra and let $h : C(X) \to A$ be a unital monomorphism. Suppose that there is a unitary $u \in A$ such that*

$$h(a)u = uh(a) \text{ for all } a \in C(X) \text{ and } \mathrm{Bott}(h, u) = 0. \tag{e 11.1}$$

Suppose also that $\psi : C(X \times S^1) \to A$ defined by $\psi(f \otimes g) = h(f)g(u)$ for $f \in C(X)$ and $g \in C(S^1)$ is a monomorphism. Then, for any $\epsilon > 0$ and any finite subset $\mathcal{F} \subset C(X)$, there exists a rectifiable continuous path of unitaries $\{u_t : t \in [0,1]\}$ of A such that

$$u_0 = u, \ u_1 = 1_A \text{ and } \|[h(a), u_t]\| < \epsilon \tag{e 11.2}$$

for all $a \in \mathcal{F}$ and all $t \in [0,1]$. Moreover,

$$\mathrm{Length}(\{u_t\}) \le \pi + \epsilon\pi. \tag{e 11.3}$$

PROOF. Let $\epsilon > 0$ and $\mathcal{F} \subset C(X)$ be a finite subset. Without loss of generality, we may assume that $1_{C(X)} \in \mathcal{F}$. Let $\mathcal{F}_1 = \{a \otimes b : a \in \mathcal{F}, b = u, b = 1\}$. Let $\eta = \sigma_{X \times S^1, \mathcal{F}, \epsilon/16}$. Let $\delta > 0$, $\mathcal{G} \subset C(X \times S^1)$ be a finite subset of $C(X \times S^1)$ and $\mathcal{P} \subset \underline{K}(C(X \times S^1))$ be a finite subset required by 10.3 associated with $\epsilon/16$ and $\mathcal{F}_1$.

We may assume that δ and $\mathcal{G}$ are so chosen that, for any two δ_1-$\mathcal{G}$-multiplicative contractive completely positive linear map $L_1, L_2 : C(X \times S^1) \to B$ (any unital C^*-algebra B),

$$[L_1]|_{\mathcal{P}} = [L_2]|_{\mathcal{P}},$$

provided that

$$L_1 \approx_\delta L_2 \text{ on } \mathcal{G}.$$

With even smaller δ we may also assume that $\mathcal{G} = \mathcal{F}_2 \otimes \mathcal{G}_1$, where $\mathcal{F}_2 \subset C(X)$ and $\mathcal{G}_1 \subset C(S^1)$ are finite subsets and $1_{C(X)} \subset \mathcal{F}_2$ and $1_{C(S^1)} \subset \mathcal{G}_1$.

Choose $x \in X$ and let $\xi = x \times 1$.

It follows from 10.9 that there is a non-zero projection $e \in A$ with $[e] = 0$ in $K_0(A)$ and a unital monomorphism $\psi' : C(X \times S^1) \to (1-e)A(1-e)$ such that

$$\|\psi(f) - (\psi'(f) + f(\xi)e)\| < \delta_1/2 \text{ for all } f \in \mathcal{G}. \tag{e 11.4}$$

Write $e = e_1+e_2$, where e_1 and e_2 are two non-zero mutually orthogonal projections with $[e_1] = [1_A]$ and $[e_2] = -[1_A]$ in $K_0(A)$.

It follows from 10.6 that there are unital monomorphisms $h_1 : C(X) \to e_1Ae_1$ and $h_2 : C(X) \to e_2Ae_2$ such that

$$[h_1] = [h] \text{ and } [h_2] = -[h] \text{ in } KL(C(X), A). \tag{e 11.5}$$

Define $\psi_1 : C(X \times S^1) \to e_1Ae_1$ by $\psi_1(f \otimes g) = h_1(f)g(1)e_1$ for all $f \in C(X)$ and $g \in C(S^1)$ and define $\psi_2 : C(X \times S^1) \to e_2Ae_2$ by $\psi_2(f \otimes g) = h_2(f)g(1)e_2$ for all $f \in C(X)$. Define $\Psi_0 : C(X \times S^1) \to (1 - e_1)A(1 - e_1)$ by $\Psi_0 = \psi_2 + \psi'$ and define $\Psi : C(X \times S^1) \to A$ by $\Psi = \psi_1 + \Psi_0$. Note that

$$\|h(f) - (f(\xi)(e_1 + e_2) + \psi'(f \otimes 1))\| < \delta_1/2 \text{ for all } f \in C(X). \tag{e 11.6}$$

Define $\phi_0 : C(X \times S^1) \to (1 - e_1)A(1 - e_1)$ by $\phi_0(f) = \sum_{j=1}^m f(x_i)q_i$ for all $f \in C(X\times S^1)$, where $\{x_1, x_2, ..., x_m\}$ are in X and $\{q_1, q_2, ..., q_m\}$ is a set of nonzero mutually orthogonal projections such that $[q_i] = 0$ in $K_0(A)$ and $\sum_{i=1}^m q_i = (1-e_1)$. This is possible since $[1 - e_1] = 0$ in $K_0(A)$. Then

$$[\phi_0] = 0 \text{ in } KL(C(X), A). \tag{e 11.7}$$

Moreover, by choosing more points and large m, we may assume that ϕ_0 is $1/2$-$\mathcal{G}$-injective.

Note that $\mathrm{Bott}(h, u) = 0$. Thus , by 2.10 and by (e 11.4), (e 11.5) and (e 11.7),

$$[\Psi_0]|_P = [\phi_0]|_{\mathcal{P}}. \tag{e 11.8}$$

It follows from 10.3 that there exists a unitary $w_1 \in (1-e_1)A(1-e_1)$ such that

$$\mathrm{ad}\, w_1 \circ \phi_0 \approx_{\epsilon/16} \Psi_0 \text{ on } \mathcal{F}_1. \tag{e 11.9}$$

By 2.10 and by (e 11.4), (e 11.5) and (e 11.7)

$$[\Psi]|_{\mathcal{P}} = [\psi]|_{\mathcal{P}}. \tag{e 11.10}$$

It follows from 10.3 that there exists a unitary $w_2 \in A$ such that

$$\mathrm{ad}\, w_2 \circ \Psi \approx_{\epsilon/16} \psi \text{ on } \mathcal{F}_1. \tag{e 11.11}$$

Let $v = \phi_0(1 \otimes z)$. In the finite dimensional commutative C^*-subalgebra $\phi_0(C(X \times S^1)$, it is easy to find a continuous rectifiable path of unitaries $\{v_t : t \in [0, 1]\}$ in $(1 - e_1)A(1 - e_1)$ such that

$$v_0 = v, \ v_1 = 1 - e_1 \text{ and } \|[\phi_0(a \otimes 1), v_t]\| = 0 \tag{e 11.12}$$

for all $t \in [0, 1]$ and

$$\mathrm{Length}(\{v_t\}) \le \pi. \tag{e 11.13}$$

Define

$$U_t = w_2^*(e_1 + w_1^*(v_t)w_1)w_2 \text{ for all } t \in [0, 1]. \tag{e 11.14}$$

Clearly $U_1 = 1_A$. Then, by (e 11.9) and (e 11.11),

$$\begin{aligned}\|U_0 - u\| &= \|\mathrm{ad}\, w_2 \circ (\psi_1(1 \otimes z) \oplus \mathrm{ad}\, w_1 \circ \phi_0(1 \otimes z)) - u\| \\ &< \epsilon/16 + \|\mathrm{ad}\, w_2 \circ (\psi_1(1 \otimes z) \oplus \Psi_0(1 \otimes z)) - u\| \\ &< \epsilon/16 + \epsilon/16 + \|\psi(1 \otimes z) - u\| = \epsilon/8.\end{aligned} \tag{e 11.15}$$

We also have, by (e 11.9), (e 11.11) and (e 11.12),

$$\|[h(a), U_t]\| = 2(\epsilon/16) + \|[\mathrm{ad}\, w_2 \circ \Psi(a \otimes 1), U_t]\| < \epsilon/8 + \epsilon/8 = \epsilon/4 \tag{e 11.16}$$

for all $a \in \mathcal{F}$. Moreover,

$$\text{Length}(\{U_t\}) \leq \pi. \tag{e 11.17}$$

By (e 11.17), (e 11.15) and (e 11.16), we obtain a continuous rectifiable path of unitaries $\{u_t : t \in [0,1]\}$ of A such that

$$u = u_0, \ \ u_1 = 1_A \ \text{ and } \ \|[h(a), u_t]\| < \epsilon \ \text{ for all } \ t \in [0,1]. \tag{e 11.18}$$

Furthermore,

$$\text{Length}(\{u_t\}) \leq \pi + \epsilon\pi. \tag{e 11.19}$$

□

Lemma 11.2. *Let X be a compact metric space and let $\mathcal{F} \subset C(X)$ be a finite subset. For any $\epsilon > 0$, there exists $\delta > 0$ and a finite subset $\mathcal{G} \subset C(X)$ satisfying the following: Suppose that A is a unital purely infinite simple C^*-algebra, that $h : C(X) \to A$ is a unital monomorphism and that $u \in A$ is a unitary such that*

$$\|[h(a), u]\| < \delta \ \textit{for all} \ \ a \in \mathcal{G}. \tag{e 11.20}$$

Then there exists a unital monomorphism $H : C(X \times S^1) \to A$ and a rectifiable continuous path of unitaries $\{u_t : t \in [0,1]\}$ such that

$$u_0 = u, \ \ u_1 = H(1 \otimes z), \tag{e 11.21}$$

$$\|H(a \otimes 1) - h(a)\| < \epsilon \ \textit{ and } \ \|[h(a), u_t]\| < \epsilon \quad \textit{for all } t \in [0,1] \tag{e 11.22}$$

and for all $a \in \mathcal{F}$. Moreover

$$\text{Length}(\{u_t\}) \leq \pi + \epsilon. \tag{e 11.23}$$

Furthermore, if, in addition to (e 11.20), *there is a finite subset $\mathcal{P} \subset \underline{K}(C(X))$,*

$$\text{Bott}(h, u)|_{\mathcal{P}} = 0,$$

then

$$[H]|_{\boldsymbol{\beta}(\underline{K}(C(X)) \cap \mathcal{P})} = 0. \tag{e 11.24}$$

Proof. Fix $\epsilon > 0$ and a finite subset $\mathcal{F} \subset C(X)$. To simplify the notation, without loss of generality, we may assume that $\mathcal{F}$ is in the unit ball of $C(X)$. Let $\delta_1 > 0$, $\mathcal{G}_1 \subset C(X \times S^1)$ and $\mathcal{P}_1 \subset \underline{K}(C(X \times S^1))$ be a finite subset which are required in 10.4 associated with $\epsilon/32$ and $\mathcal{F} \otimes S$, where $S = \{1_{C(S^1)}, z\}$. Without loss of generality, we may assume that $\mathcal{G}_1 = \mathcal{F}_1 \otimes S$, where $\mathcal{F}_1 \subset C(X)$ is a finite subset.

We may assume that $\delta_1 < \min\{1/2, \epsilon/32\}$ and $\mathcal{F} \subset \mathcal{F}_1$. Put

$$\mathcal{P}_2 = \{(\text{id} - \hat{\boldsymbol{\beta}})(x), \hat{\boldsymbol{\beta}}(x), x : x \in \mathcal{P}_1\}.$$

By choosing possibly even smaller δ_1 and larger $\mathcal{F}_1$, we may assume that

$$[L_1]|_{\mathcal{P}_2} = [L_2]|_{\mathcal{P}_2}$$

for any pair of δ_1-$\mathcal{F}_1 \otimes S$-multiplicative contractive completely positive linear maps from $C(X \otimes S^1)$ to any unital C^*-algebra B, provided that

$$L_1 \approx_{\delta_1} L_2 \ \text{ on } \ \mathcal{F}_1 \otimes S.$$

There is $\sigma > 0$ such that for any σ-dense subset $\{x_1, x_2, ..., x_n\} \subset X$ and a finite subset $\{t_1, t_2, ..., t_m\}$ such that

$$\max\{\|f(x_i)\| : i = 1, 2, ..., n\} \geq (3/4)\|f\| \tag{e 11.25}$$

for all $f \in \mathcal{F}_1$ and

$$\max\{\|g(x_i \times t_j)\| : 1 \le i \le n \text{ and } 1 \le j \le m\} \ge (3/4)\|g\| \tag{e 11.26}$$

for all $g \in \mathcal{G}_1$.

Let $\delta_2 > 0$ and $\mathcal{G}_2 \subset C(X \times S^1)$ be a finite subset required in 2.23 for $\delta_1/4$ (in place of ϵ), $\sigma/2$ (in place of σ) and $\mathcal{G}_1$ (in place of $\mathcal{F}$). Without loss of generality, we may assume that $\delta_2 < \delta_1/4$ and $\mathcal{G}_2 = \mathcal{F}_2 \otimes S$ with $\mathcal{F}_2 \subset C(X)$ being a finite subset containing $\mathcal{F}_1$.

Let $\eta > 0$ be required in 10.1 for $\delta_2/2$ (in place of ϵ), $\sigma/2$ and $\mathcal{F}_2$. We may assume that $\eta < \delta_2/2$.

Let $\delta > 0$ and $\mathcal{G} \subset C(X)$ (in place of $\mathcal{F}_1$) be a finite subset required by 10.2 for η and $\mathcal{F}_2$ and $\sigma/2$. We may assume $\mathcal{F}_2 \subset \mathcal{G}$.

Let $\mathcal{P} \subset \underline{K}(C(X))$ be a finite subset such that $\boldsymbol{\beta}(\mathcal{P}) \supset \hat{\boldsymbol{\beta}}(\mathcal{P}_1)$. Suppose that h and u satisfy the conditions in the theorem for above δ, $\mathcal{G}$ and $\mathcal{P}$. It follows from 10.2 that there is an η-$\mathcal{G}_2$-multiplicative contractive completely positive linear map $\psi : C(X \times S^1) \to A$ such that

$$\|\psi(f \otimes g) - h(f)g(u)\| < \eta < \delta_2/2 \text{ for all } f \in \mathcal{F}_2 \text{ and } g \in S. \tag{e 11.27}$$

Moreover, combining with 10.1, we also have

$$\|\psi(f \otimes g) - (\Psi_1(f) + \sum_{i=1}^{n} f(x_i)g(s_i)e_i)\| < \delta_2/2 \tag{e 11.28}$$

for all $f \in \mathcal{F}_2$ and $g \in S$, where $\Psi_1 : C(X \times S^1) \to (1-e)A(1-e)$ is a unital δ_2-$\mathcal{G}_2$-multiplicative contractive completely positive linear map, $e_1, e_2, ..., e_n$ are non-zero mutually orthogonal projections, $e = \sum_{i=1}^{n} e_i$, $[e_i] = 0$ in $K_0(A)$ $(i = 1, 2, ..., n)$, $\{x_1, x_2, ..., x_n\}$ is σ-dense, and $s_1, s_2, ..., s_n$ are points in S^1 (not necessarily distinct). Define $\Psi_1' : C(X \otimes S^1) \to eAe$ by $\Psi_1'(f) = \sum_{i=1}^{n} f(x_i \times s_i)e_i$ for all $f \in C(X \times S^1)$. It follows that

$$[\Psi_1'] = 0 \text{ in } KL(C(X), A). \tag{e 11.29}$$

There are m non-zero mutually orthogonal projections $e_{i,1}, e_{i,2}, ..., e_{i,m}$ in e_iAe_i such that $[e_{i,j}] = 0$ in $K_0(A)$, $j = 1, 2, ..., m$, for each i. Define $\Psi_0 : C(X \times S^1) \to eAe$ by $\Psi_0(f \otimes g) = \sum_{i=1}^{n}(\sum_{j=1}^{m} f(x_i)g(t_j)e_{i,j})$ for all $f \in C(X)$ and $g \in C(S^1)$.

There is a continuous path of unitaries $\{w_t : t \in [0, 1]\}$ in the finite dimensional commutative C^*-subalgebra $\Psi_0(C(X \times S^1))$ such that

$$\text{Length}(\{w_t\}) \le \pi, \quad w_0 = \sum_{i=1}^{n} s_ie_i, \quad w_1 = \Psi_0(1 \otimes z). \tag{e 11.30}$$

Moreover,

$$w_t\Psi_0(f \otimes 1) = \Psi_0(f \otimes 1)w_t \text{ for all } t \in [0, 1] \tag{e 11.31}$$

and for all $f \in C(X)$. Note also

$$[\Psi_0] = 0 \text{ in } KL(C(X \times S^1), A). \tag{e 11.32}$$

By (e 11.25), Ψ_0 is $(3/4)$-$\mathcal{G}_1$-injective. So $\Psi_1 + \Psi_0$ is δ_2-$\mathcal{G}_1$-multiplicative and $(3/4)$-$\mathcal{G}_1$-injective. Let $\Phi_0 : C(X \times S^1) \to A$ by $\Phi_0(f \otimes g) = h(f)g(1)$ for all $f \in C(X)$ and $g \in C(S^1)$. The condition that $\text{Bott}(h, u)|_{\mathcal{P}} = 0$ (see 4.1) together with the choice of δ_1 and $\mathcal{F}_1$ implies that

$$[\Phi_0]|_{\mathcal{P}_2} = [\psi]|_{\mathcal{P}_2}. \tag{e 11.33}$$

By (e 11.28), (e 11.29), (e 11.32) and (e 11.33) and the choice of δ_1 and $\mathcal{F}_1$, we have

$$[\Phi_0]|_{\mathcal{P}_2} = [\Psi_1 + \Psi_0]|_{\mathcal{P}_2}. \tag{e 11.34}$$

By applying 10.4, we obtain a unital monomorphism $H : C(X \times S^1) \to A$ such that

$$\Psi_1 + \Psi_0 \approx_{\epsilon/32} H \text{ on } \mathcal{F}. \tag{e 11.35}$$

Furthermore, from the proof of 10.4, if $\text{Bott}(h, u)|_{\mathcal{P}} = 0$ is assumed, we can also assume that

$$[H]|_{\boldsymbol{\beta}(\underline{K}(C(X))\cap\mathcal{P})} = 0. \tag{e 11.36}$$

There is a unitary $w_0 \in (1-e)A(1-e)$ such that

$$\|\Psi_1(1 \otimes z) - w_0\| < 2\delta_2 < \delta_1 < \epsilon/32. \tag{e 11.37}$$

Define $U_t = w_0 + w_t$ for $t \in [0,1]$. Then $\{U_t : t \in [0,1]\}$ is a continuous path of unitaries in A such that

$$U_0 = w_0 + \sum_{i=1}^n s_i e_i, \;\; U_1 = H(1 \otimes z) \text{ and} \tag{e 11.38}$$

$$\|[U_t, H(f \otimes 1)]\| < 2\delta_2 + \epsilon/32 < \epsilon/16 \text{ for all } t \in [0,1] \tag{e 11.39}$$

and for all $f \in C(X)$.

Moreover,

$$\text{Length}(\{U_t\}) \le \pi. \tag{e 11.40}$$

Note also we have, by (e 11.28) and (e 11.35), and by (e 11.37)

$$H(f \otimes 1) \approx_{\epsilon/16} h(f) \text{ on } \mathcal{F} \text{ and} \tag{e 11.41}$$

$$\|u - U_0\| < \epsilon/4. \tag{e 11.42}$$

We also have, by (e 11.41) and (e 11.39),

$$\|[h(f), U_t]\| < \epsilon/8. \tag{e 11.43}$$

The lemma follows if we connect u with U_0 properly. □

Theorem 11.3. *Let X be a compact metric space. For any $\epsilon > 0$ and any finite subset $\mathcal{F} \subset C(X)$, there exists $\delta > 0$, a finite subset $\mathcal{G} \subset C(X)$ and a finite subset $\mathcal{P} \subset \underline{K}(C(X))$ satisfying the following:*

Suppose that A is a unital purely infinite simple C^*-algebra and suppose that $h : C(X) \to A$ is a unital monomorphism. Suppose that there is a unitary $u \in A$ such that

$$\|[h(a), u]\| < \delta \textit{ for all } a \in \mathcal{G} \textit{ and } \text{Bott}(h, u)|_{\mathcal{P}} = 0. \tag{e 11.44}$$

Then there is a continuous rectifiable path of unitaries $\{u_t : t \in [0,1]\}$ of A such that

$$u_0 = u, \;\; u_1 = 1_A \textit{ and } \|[h(a), u_t]\| < \epsilon \text{ for all } a \in \mathcal{F} \textit{ and } t \in [0,1]. \tag{e 11.45}$$

Moreover,

$$\|u_t - u_{t'}\| \le (2\pi + \epsilon)|t - t'| \textit{ for all } t, t' \in [0,1] \textit{ and} \tag{e 11.46}$$

$$\text{Length}(\{u_t\}) \le 2\pi + \epsilon. \tag{e 11.47}$$

PROOF. Fix $\epsilon > 0$ and finite subset $\mathcal{F} \subset C(X)$. Let $\delta > 0$, $\mathcal{G} \subset C(X)$ be a finite subset and $\mathcal{P} \subset \underline{K}(C(X))$ be a finite subset required in 11.2 for $\epsilon/4$ and $\mathcal{F}$.

Suppose that h and u satisfy the conditions in the theorem. By 11.2, there is a unital monomorphism $H : C(X \times S^1) \to A$ and there is a continuous rectifiable path of unitaries $\{v_t : t \in [0,1]\}$ in A satisfying the following:

$$v_0 = u, \;\; v_1 = H(1 \otimes z), \tag{e 11.48}$$

$$\|H(f \otimes 1) - h(f)\| < \epsilon/4, \;\; \|[h(f), v_t]\| < \epsilon/4 \tag{e 11.49}$$

for all $t \in [0,1]$ and for all $f \in \mathcal{F}$,

$$\text{Length}(\{v_t\}) \le \pi + \epsilon/4 \text{ and } [H]|_{\beta(\underline{K}(C(X)))} = 0. \tag{e 11.50}$$

Put $h_1(f) = H(f \otimes 1)$ for all $f \in C(X)$. By (e 11.50),

$$\text{Bott}(h_1, v_1) = 0. \tag{e 11.51}$$

It follows from 11.1 that there is a continuous rectifiable path $\{w_t : t \in [0,1]\}$ in A such that

$$w_0 = v_1, \;\; w_1 = 1_A, \|[h_1(f), w_t]\| < \epsilon/4 \tag{e 11.52}$$

for all $t \in [0,1]$ and $f \in \mathcal{F}$, and

$$\text{Length}(\{w_t\}) \le \pi + \epsilon\pi/4. \tag{e 11.53}$$

Now define $u_t = v_{2t}$ for $t \in [0, 1/2]$ and $u_t = w_{2t-1}$ for $t \in [1/2, 1]$. Then,

$$u_0 = v_0 = u, \;\; u_1 = w_1 = 1_A.$$

By (e 11.49) and (e 11.52),

$$\|[h(f), u_t]\| < \epsilon \text{ for all } t \in [0,1] \text{ and } \text{ for all } f \in \mathcal{F}.$$

Moreover, by (e 11.50) and (e 11.53)

$$\text{length}(\{u_t\}) \le 2\pi + \epsilon.$$

To get (e 11.46), we apply 9.3 and what we have proved above. Note that we may have to choose a different δ. □

Corollary 11.4. *Let X be a finite CW complex and let $\mathcal{F} \subset C(X)$ be a finite subset. For any $\epsilon > 0$, there exists $\delta > 0$, a finite subset $\mathcal{G} \subset C(X)$ satisfying the following:*

Suppose that A is a unital purely infinite simple C^*-algebra and suppose that $h : C(X) \to A$ is a unital monomorphism. Suppose that there is a unitary $u \in A$ such that

$$\|[h(a), u]\| < \delta \text{ for all } a \in \mathcal{G} \textit{ and } \text{Bott}(h, u) = 0. \tag{e 11.54}$$

Then there is a continuous rectifiable path of unitaries $\{u_t : t \in [0,1]\}$ of A such that

$$u_0 = u, \;\; u_1 = 1_A \textit{ and } \|[h(a), u_t]\| < \epsilon \text{ for all } a \in \mathcal{F} \textit{ and } t \in [0,1]. \tag{e 11.55}$$

Moreover,

$$\text{Length}(\{u_t\}) \le 2\pi + \epsilon\pi. \tag{e 11.56}$$

Corollary 11.5. *Let X be a compact metric space and let A be a unital purely infinite simple C^*-algebra. Suppose that $h : C(X) \to A$ is a unital monomorphism and that there is a unitary $u \in A$ such that*

$$uh(f) = h(f)u \text{ for all } f \in C(X) \text{ and } \mathrm{Bott}(h, u) = 0. \tag{e 11.57}$$

Then there is a continuous rectifiable path of unitaries $\{u_t : t \in [0,1]\}$ of A such that

$$u_0 = u, \;\; u_1 = 1_A \text{ and } \|[h(a), u_t]\| < \epsilon \text{ for all } a \in \mathcal{F} \text{ and } t \in [0,1]. \tag{e 11.58}$$

Moreover,

$$\mathrm{Length}(\{u_t\}) \le 2\pi + \epsilon\pi. \tag{e 11.59}$$

Remark 11.6. As we have seen in the case that A is a unital simple C^*-algebra of real rank zero and stable rank one, one essential difference of the original Basic Homotopy Lemma from the general form is that the constant δ could no longer be universal whenever the dimension of X is at least two. In that case, a measure distribution becomes part of the statement. This additional factor disappears when A is purely infinite. (There is no measure/trace in this case!) However, there is a second difference. In both purely infinite and finite cases, when the dimension of X is at least two, we need to assume that h is a monomorphism, and when the dimension of X is no more than one, h is only assumed to be a homomorphism for the case that $K_1(A) = \{0\}$.

It is much easier to reveal the topological obstruction in this case. Just consider a monomorphism $\phi : C(S^1 \times S^1) \to A$ with $\mathrm{bott}_1(\phi) \neq 0$ and $\mathrm{bott}_0(\phi) = 0$. In other words,

$$\mathrm{bott}_1(u, v) \neq 0 \text{ and } [v] = 0,$$

where $u = \phi(z \otimes 1)$ and $v = \phi(1 \otimes z)$. This is possible if A is a unital purely infinite simple C^*-algebra or A is a unital separable simple C^*-algebra with tracial rank zero so that $\ker\rho_A \neq \{0\}$. Let D be the closed unit disk and view S^1 as a compact subset of D. Define $h : C(D) \to A$ by $h(f) = \phi(f|_{S^1}(z \otimes 1))$ for all $f \in C(D)$. Since D is contractive,

$$\mathrm{Bott}(h, v) = 0.$$

But one can not find any continuous path of unitaries $\{v_t : t \in [0,1]\}$ of A such that

$$v_0 = v, \;\; v_1 = 1_A \text{ and } \|[h(f), v_t]\| < \epsilon$$

for all $t \in [0,1]$ and for some small ϵ, where $f(z) = z$ for all $z \in D$, since $\mathrm{bott}_1(u, v) \neq 0$ and $h(f) = u$. Counterexamples for the case that A is a unital separable simple C^*-algebra with tracial rank zero for which $\ker\rho_A = \{0\}$ can also be made but not for commuting pairs. One can begin with a sequence of unitaries $\{u_n\}$ and $\{v_n\}$ with $v_n \in U_0(A)$ such that $\lim_{n\to\infty} \|[u_n, v_n]\| = 0$ and $\mathrm{bott}_1(u_n, v_n) \neq 0$. Then define $h_n : C(D) \to A$ by $h_n(f) = f(u_n)$ for $f \in C(D)$. Again $\mathrm{Bott}(h_n, v_n) = 0$ and $\lim_{n\to\infty} \|[h_n(f), v_n]\| = 0$ for all $f \in C(D)$. This will give a counterexample. In summary, the condition that h is monomorphism can not be replaced by the condition that h is a homomorphism in the Basic Homotopy Lemma whenever $\dim X \geq 2$.

CHAPTER 4

Approximate homotopy

12. Homotopy length

In the next two sections, we will study when two maps from $C(X)$ to A, a unital purely infinite simple C^*-algebra, or a separable simple C^*-algebra of tracial rank zero, are approximately homotopic. When two maps are approximately homotopic, we will give an estimate for the length of the homotopy. In this section we will discuss the notion of the length of a homotopy.

Recall that $\mathbf{B}$ denotes the class of unital C^*-algebras which are simple C^*-algebras with real rank zero and stable rank one, or purely infinite simple C^*-algebras.

Lemma 12.1. *Let $A \in \mathbf{B}$ be a unital simple C^*-algebra and let X be a compact metric space. Let $\epsilon > 0$ and $\mathcal{F} \subset C(X)$ be a finite subset. Suppose that $h_0, h_1 : C(X) \to A$ are two unital homomorphisms such that there is unitary $u \in A$ such that*

$$\operatorname{ad} u \circ h_1(f) \approx_{\epsilon/2} h_0(f) \quad \text{on } \mathcal{F}.$$

Then, there is a unitary $v \in U_0(A)$ such that

$$\operatorname{ad} v \circ h_1 \approx_{\epsilon} h_0 \quad \text{on } \mathcal{F}.$$

PROOF. Fix ϵ and $\mathcal{F} \subset C(X)$. Let $\eta = \sigma_{X,\mathcal{F},\epsilon/2}$. For a point $\xi \in X$, let $g \in C(X)_+$ with support in $O(\xi, \eta) = \{x \in X : \operatorname{dist}(x, \xi) < \eta\}$. We may choose such ξ in the spectrum of h_0 so that $h_0(g) \neq 0$. Then there is a non-zero projection $e \in \overline{h_0(g)Ah_0(g)}$.

Since A is assumed to be purely infinite or has real rank zero and stable rank one, there is a unitary $w \in eAe$ so that $[(1-e)+w] = [u^*]$ in $K_1(A)$. Moreover, $u((1-e)+w) \in U_0(A)$. Put $v = u((1-e)+w)$. Note that, by the choice of η,

$$\|((1-e)+w)^* h_0(f) - h_0(f)((1-e)+w)\| < \epsilon/2$$

for all $f \in \mathcal{F}$. Thus

$$\operatorname{ad} v \circ h_1 \approx_{\epsilon} h_0 \text{ on } \mathcal{F}.$$

□

Definition 12.2. Let A and B be unital C^*-algebras. Let $H : A \to C([0,1], B)$ be a contractive completely positive linear map. One has the following notion of the length of the homotopy. For any partition $P : 0 = t_0 < t_1 < \cdots < t_m = 1$ and $f \in A$, put

$$L_H(P, f) = \sum_{i=1}^{m} \|\pi_{t_i} \circ H(f) - \pi_{t_{i-1}} \circ H(f)\| \text{ and}$$
$$L_H(f) = \sup_P L_H(P, f).$$

Define

$$\text{Length}(\{\pi_t \circ H\}) = \sup_{f\in A, \|f\|\le 1} L_H(f).$$

Theorem 12.3. *Let A be a separable unital simple C^*-algebra of tracial rank zero and let X be a compact metric space. Suppose that $h_0, h_1 : C(X) \to A$ are two unital monomorphisms such that*

$$[h_0] = [h_1] \ \text{in } KL(C(X), A) \text{ and} \tag{e 12.1}$$

$$\tau \circ h_1 = \tau \circ h_2 \ \text{ for all } \ \tau \in T(A). \tag{e 12.2}$$

Then, for any $\epsilon > 0$ and any finite subset $\mathcal{F} \subset C(X)$, there is a unital homomorphism $\Phi : C(X) \to C([0,1], A)$ such that $\pi_t \circ H$ is a unital monomorphism for all $t \in [0,1]$,

$$h_0 \approx_\epsilon \pi_0 \circ \Phi \ \text{ on } \mathcal{F} \text{ and } \pi_1 \circ \Phi = h_1. \tag{e 12.3}$$

Moreover,

$$\text{Length}(\{\pi_t \circ \Phi\}) \le 2\pi. \tag{e 12.4}$$

Proof. It follows from Theorem 3.3 of [**33**] that there exists a sequence of unitaries $w \in A$ such that

$$\lim_{n\to\infty} \text{ad}\, w_n \circ h_1(f) = h_0(f) \ \text{ for all } \ f \in C(X).$$

It follows from 12.1 that there is $z \in U_0(A)$ such that

$$\text{ad}\, z \circ h_1 \approx_{\epsilon/2} h_0 \ \text{ on } \ \mathcal{F}.$$

It follows from [**21**] that, for any $\epsilon > 0$, there is a continuous path of unitaries $\{z_t : t \in [0,1]\}$ in A such that

$$z_0 = 1, \ \ \|z_1 - z\| < \epsilon/2M,$$

where $M = \max\{\|f\| : f \in \mathcal{F}\}$, and

$$\text{Length}(\{z_t\}) \le \pi.$$

Define $H : C(X) \to C([0,1], A)$ by

$$\pi_t \circ H(f) = \text{ad}\, z_t \circ h_1.$$

Then

$$\text{Length}(\{\pi_t \circ H\}) \le 2\pi.$$

□

12.4. Let $\xi_1, \xi_2 \in X$. Suppose that there is a continuous rectifiable path $\gamma : [0,1] \to X$ such that $\gamma(0) = \xi_1$ and $\gamma(1) = \xi_2$. Let A be a unital C^*-algebra. Define $\psi_1, \psi_2 : C(X) \to A$ by $\psi_i(f) = f(\xi_i) \cdot 1_A$. These two homomorphisms are homotopic. In fact one can define $H : C(X) \to C([0,1], A)$ by $\pi_t \circ H(f) = f(\gamma(t))$ for all $f \in C(X)$. However, with definition 12.2, $\text{Length}(\{\pi_t \circ H\}) = \infty$ even in the case that $X = [0,1]$, $\xi_1 = 0$ and $\xi_2 = 1$. This is because the presence of continuous functions of unbounded variation. Nevertheless $\{\pi_t \circ H\}$ should be regarded a good homotopy. This leads us to consider a different notion of the length.

Definition 12.5. Let X be a path connected compact metric space and fix a base point ξ_X. Denote by $P_t(x,y)$ a continuous path in X which starts at x and ends at y. By *universal homotopy length*, we mean the following constant:

$$L(X,\xi_X) = \sup_{y\in X} \inf\{\mathrm{Length}(\{P_t(\xi_X,y)\}) : P_t(\xi_X,y)\}. \tag{e 12.5}$$

Define

$$\underline{L}_p(X) = \inf_{\xi_X\in X} L(X,\xi_X) \text{ and } \bar{L}_p(X) = \sup_{\xi_X\in X} L(X,\xi_X).$$

The *total length* of X, denote by $L(X)$, is the following constant:

$$L(X) = \sup_{\xi,\zeta\in X} \inf\{\mathrm{Length}(\{P_t(\xi,\zeta)\} : P_t(\xi,\zeta)\}. \tag{e 12.6}$$

It is clear that

$$\underline{L}_p(X) \le L(X,\xi_X) \le \bar{L}_p(X) \le L(X) \le 2L(X,\xi_X). \tag{e 12.7}$$

If D is the closed unit disk, then

$$\underline{L}_p(D) = L(D,\{0\}) = 1 \text{ and } \bar{L}_p(D) = L(D) = 2. \tag{e 12.8}$$

If S^1 is the unit circle, then

$$\underline{L}_p(S^1,1) = L(S^1,1) = \bar{L}_p(S^1) = L(S^1) = \pi. \tag{e 12.9}$$

There are plenty of examples that $L(X,\xi_X) = \infty$ for any $\xi_X \in X$. For example, let X be the closure of the image of map $f(t) = t\sin(1/t)$ for $t \in (0,1]$.

Definition 12.6. Let X be a path connected compact metric space. We fix a metric $d(-,-) : X\times X \to \mathbb{R}_+$. A function $f \in C(X)$ is sad to be Lipschitz if

$$\sup_{x,x'\in X, x\neq x'} \frac{|f(x)-f(x')|}{d(x,x')} < \infty.$$

For Lipschitz function f, define

$$\bar{D}_f = \sup_{x\in X} \lim_{\delta\to 0} \sup_{x',x''\in O(x,\delta)} \frac{|f(x')-f(x'')|}{d(x',x'')}. \tag{e 12.10}$$

Note that

$$\bar{D}_f \le \sup_{x,x'\in X, x\neq x'} \frac{|f(x)-f(x')|}{d(x,x')} < \infty.$$

Definition 12.7. Let A be a unital C^*-algebra and let $H : C(X) \to C([0,1],A)$ be a homotopy path from $\pi_0\circ H$ to $\pi_1\circ H$.

By the homotopy length of $\{\pi_t\circ H\}$, denote by

$$\overline{\mathrm{Length}}(\{\pi_t\circ H\}),$$

we mean

$$\sup\{L_H(f) : f\in C(X), \bar{D}_f \le 1\}.$$

Proposition 12.8. *Let X be a connected compact metric space. For any unital C^*-algebra A and unital homomorphism $H : C(X) \to C([0,1],A)$,*

$$\overline{\mathrm{Length}}(\{\pi_t\circ H\}) \le \underline{L}_p(X)\mathrm{Length}(\{\pi_t\circ H\}) \tag{e 12.11}$$

for any $\xi_X \in X$.

PROOF. Note that if $\mathrm{Length}(\{\pi_t \circ H\}) = 0$, then $\overline{\mathrm{Length}}(\{\pi_t \circ H\}) = 0$. So we may assume that $\mathrm{Length}(\{\pi_t \circ H\}) \neq 0$. In this case, it is clear that we may assume that $\underline{L}_p(X) < \infty$. It follows that $L(X) < \infty$ (by (e 12.7)).

Fix $\xi_X \in X$. We first assume that

$$\overline{\mathrm{Length}}(\{\pi_t \circ H\}) < \infty.$$

Fix $\epsilon > 0$. There is a Lipschitz function $f \in C(X)$ with $\bar{D}_f \leq 1$ such that

$$L_H(f) > \overline{\mathrm{Length}}(\{\pi_t \circ H\}) - \epsilon. \tag{e 12.12}$$

Fix $x \in X$. Let $\eta > 0$. There is a continuous rectifiable path $Q_t(x, \xi_X)$ which connects x with ξ_X such that

$$\mathrm{Length}(\{Q_t(x,\xi_X)\}) < \inf\{\mathrm{Length}(\{P_t(x,\xi_X)\}) : P_t(x,\xi_X)\} + \eta. \tag{e 12.13}$$

Denote by Q^* the image of the path $\{Q_t : t \in [0,1]\}$. It is a compact subset of X. For each $y \in Q^*$, there is a $\delta_y > 0$ such that

$$|f(y') - f(y'')| \leq (1+\eta) d(y, y') \tag{e 12.14}$$

for any $y', y'' \in O(y, \delta_y)$ (Note that $\bar{D}_f \leq 1$). Since $\cup_{y \in Q^*} O(y, \delta_y/2) \supset Q^*$, one easily obtains a partition $\{t_0 = 0 < t_1 < t_2 < \cdots < t_n = 1\}$ such that, if $x \in [t_{i-1}, t_i]$, then

$$|f(x) - f(Q_{t_i})| \leq (1+\eta) d(x, Q_{t_i}) \text{ and} \tag{e 12.15}$$

$$|f(Q_{t_i}) - f(Q_{t_{i-1}})| \leq (1+\eta) d(Q_{t_i-1}, Q_{t_i}), \tag{e 12.16}$$

$i = 1, 2, ..., n$.

Put $g = f(x) - f(\xi_X)$. Then

$$\begin{aligned} |g(x)| &= |f(x) - f(\xi_X)| \leq \sum_{i=1}^n |f(Q_{t_i}) - f(Q_{t_{i-1}})| \\ &\leq (1+\eta)\sum_{i=1}^n d(Q_{t_i}, Q_{t_{i-1}}) \\ &\leq (1+\eta)\mathrm{Length}(\{Q_t(x,\xi)\}) \\ &\leq (1+\eta) L(X, \xi_X) + \eta + \eta^2 \end{aligned} \tag{e 12.17}$$

for all $\eta > 0$. It follows that $\|g\| \leq L(X, \xi_X)$. Note that

$$\pi_t \circ H(g) = \pi_t \circ H(f) - f(\xi_X) \cdot 1_A. \tag{e 12.18}$$

It follows from (e 12.18) and (e 12.12) that

$$L_H(g) = L_H(f) > \overline{\mathrm{Length}}(\{\pi_t \circ H\}) - \epsilon. \tag{e 12.19}$$

Therefore

$$L(X,\xi_X)\mathrm{Length}(\{\pi_t \circ H\}) \geq L_H(g) > \overline{\mathrm{Length}}(\{\pi_t \circ H\}) - \epsilon \tag{e 12.20}$$

(e 12.21)

for all $\epsilon > 0$. Let $\epsilon \to 0$. We then obtain

$$\overline{\mathrm{Length}}(\{\pi_t \circ H\}) \leq L(X,\xi_X)\mathrm{Length}(\{\pi_t \circ H\}) \tag{e 12.22}$$

for any $\xi_X \in X$. Thus (e 12.11) holds.

If $\overline{\mathrm{Length}}(\{\pi_t \circ H\}) = \infty$, let $L > 0$. We replace (e 12.12 by

$$L_H(f) > L. \tag{e 12.23}$$

Then, instead of (e 12.19), we obtain that

$$L_H(g) = L_H(f) > L. \tag{e 12.24}$$

Therefore

$$L(X, \xi_X)\text{Length}(\{\pi_t \circ H\}) \ge L_H(g) > L. \tag{e 12.25}$$

It follows that

$$\text{Length}(\{\pi_t \circ H\}) = \infty.$$

□

Fix a base point ξ_X and a point $x \in X$. Let $\gamma : [0, 1] \to X$ be a continuous path such that $\gamma(0) = x$ and $\gamma(1) = \xi_X$. Let A be a unital C^*-algebra. Define $h_0(f) = f(x)1_A$ and $h_1(f) = f(\xi_X)1_A$ for all $f \in C(X)$. One obtains a homotopy path $H : C(X) \to C([0, 1], A)$ by $\pi_t \circ H(f) = f(\gamma(t))1_A$ for $f \in C(X)$. Then one has the following easy fact:

Proposition 12.9.

$$\overline{\text{Length}}(\{\pi_t \circ H\}) \le \text{Length}(\{\gamma(t) : t \in [0, 1]\}). \tag{e 12.26}$$

PROOF. Let $\epsilon > 0$. There is $f \in C(X)$ with $\bar{L}_f \le 1$ such that

$$L_H(f) > \overline{\text{Length}}(\{\pi_t \circ H\}) - \epsilon/2. \tag{e 12.27}$$

There is a partition P :

$$0 = t_0 < t_1 < \cdots < t_n = 1$$

such that

$$\sum_{j=1}^{n} |f(\gamma(t_j)) - f(\gamma(t_{j-1}))| > L_H(f) - \epsilon/2. \tag{e 12.28}$$

Let $\eta > 0$. For each $t \in [0, 1]$, there is $\delta_t > 0$ such that

$$|f(\gamma(t')) - f(\gamma(t''))| \le (1 + \eta)d(\gamma(t'), \gamma(t'')) \tag{e 12.29}$$

if $\gamma(t'), \gamma(t'') \in O(\gamma(t), \delta_t)$. From this, one obtains a finer partition

$$0 = s_0 < s_1 < \cdots < s_m = 1$$

such that

$$|f(\gamma(s_i)) - f(\gamma(s_{i-1}))| \;\le\; (1 + \eta)d(\gamma(s_i), \gamma(s_{i-1})), \tag{e 12.30}$$

$i = 1, 2, ..., n$ and

$$\sum_{i=1}^{m} |f(\gamma(s_i)) - f(\gamma(s_{i-1}))| \;>\; L_H(f) - \epsilon/2. \tag{e 12.31}$$

It follows that

$$\begin{aligned} L_H(f) - \epsilon/2 \;&<\; \sum_{i=1}^{m} (1 + \eta)d(\gamma(s_i), \gamma(s_{i-1}) \\ &\le\; (1 + \eta)\text{Length}(\{\gamma(t) : t \in [0, 1]\}) \end{aligned} \tag{e 12.32}$$

for any $\eta > 0$. Let $\eta \to 0$, one obtains

$$L_H(f) - \epsilon/2 \le \text{Length}(\{\gamma(t) : t \in [0, 1]\}). \tag{e 12.33}$$

Combining this with (e 12.27), one has

$$\overline{\text{Length}}(\{\pi_t \circ H\}) - \epsilon \le \text{Length}(\{\gamma(t) : t \in [0, 1]\}). \tag{e 12.34}$$

Let $\epsilon \to 0$. One conclude that

$$\overline{\mathrm{Length}}(\{\pi_t \circ H\}) \le \mathrm{Length}(\{\gamma(t) : t \in [0,1]\}). \tag{e 12.35}$$

□

Remark 12.10. If X is a path connected compact subset of $\mathbb{C}^n$, with

$$\mathrm{dist}(\xi, \zeta) = \max_{1 \le i \le n} |x_i - y_i|,$$

where $\xi = (x_1, x_2, ..., x_n)$ and $y = (y_1, y_2, ..., y_n)$. One easily shows that (e 12.26) becomes an equality. When $X = \gamma$ and γ is smooth, then it is also easy to check that (e 12.26) becomes equality.

Lemma 12.11. *Let X be a path connected compact metric space and let $\xi_X \in X$ be a point.*

(1) *Then $L(X \times S^1, \xi_X \times 1) \le \sqrt{L(X, \xi)^2 + \pi^2}$ for all $\xi \in X$ (see 2.14).*

(2) *Suppose that A is a unital C^*-algebra and $h : C(X) \to A$ is unital homomorphism with finite dimensional range. Let $\psi : C(X) \to A$ be defined by $\psi(f) = f(\xi_X) \cdot 1_A$. Then, for any $\epsilon > 0$, and finite subset $\mathcal{F} \subset C(X)$, there exist two homomorphisms $H_1, H_2 : C(X) \to C([0,1], A)$ such that*

$$\pi_0 \circ H_1 = h, \ \pi_1 \circ H_1 = \psi \text{ for all } f \in C(X), \tag{e 12.36}$$
$$\overline{\mathrm{Length}}(\{\pi_t \circ H_1\}) \le L(X, \xi_X) + \epsilon \tag{e 12.37}$$

and

$$\pi_0 \circ H_2 = h, \ \pi_1 \circ H_2 \approx_\epsilon \psi \text{ for all } f \in \mathcal{F} \text{ and} \tag{e 12.38}$$
$$\overline{\mathrm{Length}}(\{\pi_t \circ H_2\}) \le L(X, \xi_X). \tag{e 12.39}$$

(3) *Suppose that $h_1, h_2 : C(X) \to A$ are two unital homomorphisms so that $h_1(C(X)), h_2(C(X)) \subset B \subset A$, where B is a unital commutative finite dimensional C^*-subalgebra. Then, for any $\epsilon > 0$, there is a unital homomorphism $H : C(X) \to C([0,1], B)$ such that*

$$\pi_t \circ H = h_1, \ \pi_1 \circ H = h_2 \text{ and} \tag{e 12.40}$$
$$\overline{\mathrm{Length}}(\{\pi_t \circ H\}) \le L(X) + \epsilon. \tag{e 12.41}$$

(4) *Suppose that $h : C(X \times S^1) \to A$ is a unital homomorphism with finite dimensional range, where A is a unital C^*-algebra. Then, for any $\epsilon > 0$, there exists a homomorphism $H : C(X \times S^1) \to C([0,1], A)$ such that*

$$\pi_0 \circ H = h, \ \pi_1 \circ H(f) = f(\xi_X \times 1)1_A \text{ for all } f \in C(X \times S^1), \tag{e 12.42}$$
$$\overline{\mathrm{Length}}(\{\pi_t \circ H|_{C(X)\otimes 1}\}) \le L(X, \xi_X) + \epsilon, \tag{e 12.43}$$
$$\overline{\mathrm{Length}}(\{\pi_t \circ H|_{1 \otimes C(S^1)}\}) \le \pi \text{ and} \tag{e 12.44}$$
$$[\pi_t \circ H(1 \otimes z), \pi_{t'} \circ H(f \otimes 1)] = 0 \tag{e 12.45}$$

for any $f \in C(X)$ and $t, t' \in [0,1]$.

(5) *If $h_1, h_2 : C(X \times S^1) \to B$ are two unital homomorphisms, where B is a unital commutative finite dimensional C^*-algebra, then for any $\epsilon > 0$, there exists*

a homomorphism $H : C(X \times S^1) \to C([0,1], B)$ *such that*

$$\pi_0 \circ H = h_1, \ \pi_1 \circ H = h_2 \tag{e 12.46}$$

$$\overline{\mathrm{Length}}(\{\pi_t \circ H|_{C(X)\otimes 1}\}) \le L(X) + \epsilon, \tag{e 12.47}$$

$$\overline{\mathrm{Length}}(\{\pi_t \circ H|_{1\otimes C(S^1)}\}) \le \pi. \tag{e 12.48}$$

PROOF. We may assume that $L(X, \xi_X) < \infty$. Then $L(X) < \infty$.

(1) is obvious.

For (2), let

$$h(f) = \sum_{i=1}^n f(x_i)e_i \ \text{ for all } \ f \in C(X),$$

where $x_1, x_2, ..., x_n \in X$ and $\{e_i : i = 1, 2, ..., n\}$ is a set of mutually orthogonal projections such that $\sum_{i=1}^n e_i = 1_A$. For any $\epsilon > 0$, define

$$\eta = \sigma_{X,\mathcal{F},\epsilon} \ \text{and} \ \epsilon_1 = \min\{\epsilon, \eta\}.$$

For each i, there is a rectifiable continuous path $\gamma_i : [0,1] \to X$ such that $\gamma_i(0) = x_i$ and $\gamma_i(1) = \xi_X$ such that

$$\mathrm{Length}(\{\gamma_i\}) < L(X, \xi_X) + \epsilon_1, \ \ i = 1, 2, ..., n,$$

Define $H_1 : C(X \times S^1) \to C([0,1], A)$ by

$$\pi_t \circ H_1(f) = \sum_{i=1}^n f(\gamma_i(t))e_i$$

for all $f \in C(X \times S^1)$ and $t \in [0,1]$.

Fix a partition $\mathcal{P} : 0 = t_0 < t_1 < \cdots < t_k = 1$. For any $f \in C(X)$ with $\bar{L}_f \le 1$, by applying 12.9,

$$\begin{aligned}
\sum_{j=1}^k \|\pi_{t_j} \circ H_1(f) - \pi_{t_{j-1}} \circ H_1(f)\| &= \sum_{j=1}^k (\sum_{i=1}^n |f(\gamma_i(t_j)) - f(\gamma_i(t_{j-1}))| e_i) \\
&= \sum_{i=1}^n \sum_{j=1}^k |f(\gamma_i(t_j)) - f(\gamma_i(t_{j-1}))| e_i) \\
&\le \max_{1\le i\le n} \{\sum_{j=1}^k |f(\gamma_i(t_j)) - f(\gamma_i(t_{j-1}))|\} \\
&\le L(X, \xi_X) + \epsilon_1. \qquad \text{(e 12.49)}
\end{aligned}$$

It follows that

$$\overline{\mathrm{Length}}(\{\pi_t \circ H_1\}) \le L(X, \xi_X) + \epsilon. \tag{e 12.50}$$

By re-parameterizing the paths γ_i, we may assume that, for some $a \in (0,1)$,

$$\begin{aligned}
&\mathrm{Length}(\{\gamma_i(t) : t \in [0,a]\}) \le L(X, \xi_X) \ \text{ and} \\
&\mathrm{Length}(\{\gamma_i(t) : t \in [a,1]\}) < \eta, \qquad \text{(e 12.51)}
\end{aligned}$$

$i = 1, 2, ..., n$. Define $H_2 : C(X) \to C([0,1], A)$ by $\pi_t \circ H_2 = \pi_{\frac{t}{a}} \circ H_1$. We have that

$$\mathrm{dist}(\xi_X, \gamma_i(a)) < \eta, \ \ i = 1, 2, ..., n.$$

It follows that, for $f \in \mathcal{F}$,

$$\begin{aligned} \|\pi_1 \circ H_2(f) - \psi(f)\| &= \|\pi_a \circ H_1(f) - f(\xi_X)\cdot 1_A\| \\ &= \|\sum_{i=1}^n f(\gamma_i(a))e_i - \sum_{i=1}^n f(\xi_X)e_i\| < \epsilon. \end{aligned} \tag{e 12.52}$$

Moreover,

$$\overline{\text{Length}}(\{\pi_t \circ H_2\}) \le L(X, \xi_X). \tag{e 12.53}$$

For (3), suppose that B is the commutative finite dimensional C^*-subalgebra generated by $\{e_1, e_2, ..., e_m\}$, where $e_1, e_2, ..., e_m$ are non-zero mutually orthogonal projections. We may write

$$h_i(f) = \sum_{k=1}^m f(x_{i,k})e_k \text{ for all } f \in C(X),$$

where $\{x_{i,k} : 1 \le k \le m\}$ is a set of (not necessarily distinct) points in X, $i = 1, 2$. For any $\epsilon > 0$, for each k, there exists a continuous path $\{\gamma_k : t \in [0,1]\}$ in X such that

$$\gamma_k(0) = x_{1,k}, \ \ \gamma_k(1) = x_{2,k} \text{ and}$$
$$\text{Length}(\{\gamma_k(t)\}) \le L(X) + \epsilon.$$

Define $H : C(X) \to C([0,1], B)$ by

$$\pi_t \circ H(f) = \sum_{k=1}^m f(\gamma_k(t))e_k \text{ for all } f \in C(X).$$

Similar to the estimate (e 12.49), we obtain that

$$\overline{\text{Length}}(\{\pi_t \circ H\}) \le L(X) + \epsilon.$$

For (4), we write

$$h(f) = \sum_{i=1}^n \sum_{j=1}^m f(\xi_i \times t_j)e_{i,j}$$

for all $f \in C(X \times S^1)$, where $x_1, x_2, ..., x_n \in X$ and $t_1, t_2, ..., t_m \in S^1$, and $\{e_{i,j}, i = 1, 2, ..., n, j = 1, 2, ..., m\}$ is a set of mutually orthogonal projections such that $\sum_{i,j} e_{i,j} = 1_A$. For each i, there is a continuous path $\gamma_i : [0, 1/2] \to X$ such that $\gamma_i(0) = x_i$ and $\gamma_i(1) = \xi_X$. Define $h_t : C(X \times S^1) \to A$ by

$$h_t(f) = \sum_{i=1}^n \sum_{j=1}^m f(\gamma_i(t) \times t_j)e_{i,j}$$

for all $f \in C(X \times S^1)$ and $t \in [0, 1/2]$. If $L(X, \xi_X) < \infty$, then, for any $\epsilon > 0$, we may also assume that

$$\text{Length}(\{\gamma_i\}) \le L(X, \xi_X) + \epsilon, \ \ i = 1, 2, ..., n.$$

There is for each j, a rectifiable continuous path $\lambda_j : [1/2, 1] \to S^1$ such that

$$\lambda_j(0) = t_j, \ \ \lambda_j(1) = 1 \text{ and } \text{Length}(\{\lambda_j\}) \le \pi$$

$j = 1, 2, ..., m$. For $t \in [1/2, 1]$, define

$$h_t(f) = \sum_{i=1}^n \sum_{j=1}^m f(\xi_X \times \lambda(t))e_{i,j}$$

for all $f \in C(X \times S^1)$.

Note that

(e 12.54) $$h_t(f)h_{t'}(f) = h_{t'}(f)h_t(f) \text{ for all } f \in C(X \times S^1)$$

and any $t, t' \in [0, 1]$. We then define $H : C(X \times S^1) \to C([0, 1], A)$ by

$$H(f)(t) = h_t(f) \text{ for all } f \in C(X) \text{ and } t \in [0, 1].$$

By the proof of (2), it is easy to check that (e 12.42), (e 12.43),(e 12.44) and (e 12.45) hold.

Part (5) follows from the combination of the proof of (3) and (4). □

13. Approximate homotopy for homomorphisms

Theorem 13.1. *Let X be a path connected compact metric space with the base point ξ_X and let A be a unital separable simple C^*-algebra which has tracial rank zero, or is purely infinite. Suppose that $h : C(X) \to A$ is a unital monomorphism such that*

(e 13.1) $$[h|_{C_0(Y_X)}] = 0 \text{ in } KL(C(X), A),$$

where $Y_X = X \setminus \{\xi_X\}$. Then, for any $\epsilon > 0$ and any compact subset $\mathcal{F} \subset C(X)$, there is homomorphism $H : C(X) \to C([0, 1], A)$ such that

(e 13.2) $$\pi_0 \circ H \approx_\epsilon h \text{ on } \mathcal{F} \text{ and } \pi_1 \circ H = \psi,$$

where $\psi(f) = f(\xi_X) \cdot 1_A$. Moreover,

(e 13.3) $$\overline{\text{Length}}(\{\pi_t \circ H\}) \le L(X, \xi_X).$$

PROOF. Note that $[\psi|_{C_0(Y_X)}] = 0$. As in the proof of Theorem 3.9 of [**15**], $[\psi] = [h]$ in $KL(C(X), A)$. By Theorem 3.8 of [**15**] (in the case that $TR(A) = 0$), or by Theorem 1.7 of [**6**] (in the case that A is purely infinite), there is a homomorphism $h_0 : C(X) \to A$ with finite dimensional range such that

$$h_1 \approx_{\epsilon/2} h_0 \text{ on } \mathcal{F}.$$

Since h_0 has finite dimensional range, by 12.11, there is a homomorphism $H : C(X) \to C([0, 1], A)$ such that

$$\pi_0 \circ H \approx_{\epsilon/2} h_0 \text{ on } \mathcal{F} \text{ and } \pi_1 \circ H = \psi.$$

Moreover,

$$\overline{\text{Length}}(\{\pi_t \circ H\}) \le L(X, \xi_X).$$

□

Proposition 13.2. *Let A and B be two unital C^*-algebras and let $h_1, h_2 : A \to B$ be two unital homomorphisms. Suppose that, for any $\epsilon > 0$ and any finite subset $\mathcal{F} \subset A$, there is a unital homomorphism $H : A \to C([0, 1], B)$ such that*

$$\pi_0 \circ H = h_1 \text{ and } \pi_1 \circ H \approx_\epsilon h_2 \text{ on } \mathcal{F}.$$

Then

$$[h_1] = [h_2] \text{ in } KL(A, B).$$

PROOF. Let $\mathcal{P} \subset \underline{K}(A)$ be a finite subset. Then, by 2.7, with sufficiently small ϵ and sufficiently large $\mathcal{G} \subset A$,

$$[h_1]|_{\mathcal{P}} = [\pi_t \circ H]|_{\mathcal{P}} = [h_2]|_{\mathcal{P}}.$$

It follows that $[h_1] = [h_2]$ in $KL(A, B)$. □

Proposition 13.2 states an obvious fact that if h_1 and h_2 are homotopic then $[h_1] = [h_2]$ in $KL(A, B)$. We will show that, when $A = C(X)$ and B is a unital purely infinite simple C^*-algebra or B is a unital separable simple C^*-algebra of tracial rank zero, at least for monomorphisms, $[h]$ is a complete approximately homotopy invariant.

Theorem 13.3. *Let X be a compact metric space and let A be a unital purely infinite simple C^*-algebra. Suppose that $h_1, h_2 : C(X) \to A$ are two unital monomorphisms such that*

$$[h_1] = [h_2] \text{ in } KL(C(X), A). \tag{e 13.4}$$

Then, for any $\epsilon > 0$ and any finite subset $\mathcal{F} \subset C(X)$, there is a unital homomorphism $H : C(X) \to C([0,1], A)$ such that

$$\pi_0 \circ H = h_1 \text{ and } \pi_1 \circ H \approx_\epsilon h_2 \text{ on } \mathcal{F}. \tag{e 13.5}$$

Moreover, each $\pi_t \circ H$ is a monomorphism (for $t \in [0,1]$) and

$$\mathrm{Length}(\{\pi_t \circ H\}) \le 2\pi + \epsilon. \tag{e 13.6}$$

If, in addition, X is path connected, then

$$\overline{\mathrm{Length}}(\{\pi_t \circ H\}) \le 2\pi \underline{L}_p(X) + \epsilon. \tag{e 13.7}$$

Furthermore, the converse also holds.

Proof. It follows from a theorem of Dadarlat [**6**] (see 10.3) that h_1 and h_2 are approximately unitarily equivalent, by applying 12.1, we obtain a unitary $u \in A$ with $[u] = 0$ in $K_1(A)$ such that

$$\mathrm{ad}\, u \circ h_1 \approx_\epsilon h_2.$$

By a theorem of N. C. Phillips ([**46**]), there is a continuous path of unitaries $\{u_t : t \in [0,1]\}$ (see 13.1, or see Lemma 4.4.1 of [**29**] for the exact statement) such that

$$u_0 = 1, \ \ u_1 = u \text{ and } \mathrm{Length}(\{u_t\}) \le \pi + \epsilon/2.$$

Define $H : C(X) \to C([0,1], A)$ by $H(f)(t) = \mathrm{ad}\, u_t \circ h_1(f)$ for $f \in C(X)$. Then,

$$\pi_0 \circ H = h_1 \text{ and } \pi_1 \circ H = \mathrm{ad}\, u \circ h_1.$$

Moreover, we compute that

$$\mathrm{Length}(\{\pi_t \circ H\}) \le 2\pi + \epsilon. \tag{e 13.8}$$

If X is path connected and $L(X, \xi_X) < \infty$, we may choose $\{u_t : t \in [0,1]\}$ so that

$$\mathrm{Length}(\{u_t\}) \le \pi + \frac{\epsilon}{4(1 + L(X))}.$$

Then (e 13.8) becomes

$$\mathrm{Length}(\{\pi_t \circ H\}) \le 2\pi + \frac{\epsilon}{2(1 + L(X))}.$$

By 12.8,

$$\overline{\mathrm{Length}}(\{\pi_t \circ H\}) \le 2\pi \underline{L}_p(X) + \epsilon.$$

□

Remark 13.4. The assumption that both h_1 and h_2 are monomorphisms is important. Suppose that X has at least two path connected components, say X_1 and X_2. Fix two points, $\xi_i \in X_i$, $i = 1, 2$. Suppose that A is a unital purely infinite simple C^*-algebra with $[1_A] = 0$ in $K_0(A)$. Define $h_i : C(X) \to A$ by $h_i(f) = f(\xi_i) \cdot 1_A$ for all $f \in C(X)$, $i = 1, 2$. Then $[h_1] = [h_2] = 0$ in $KL(C(X), A)$. It is clear that they are not approximately homotopic. However, when X is path connected, we have the following approximate homotopy result.

Theorem 13.5. *Let X be a path connected compact metric space and let A be a unital purely infinite simple C^*-algebra. Suppose that $h_1, h_2 : C(X) \to A$ are two unital homomorphisms such that*

$$[h_1] = [h_2] \ \ in \ \ KL(C(X), A).$$

Then, for any $\epsilon > 0$ and any finite subset $\mathcal{F} \subset C(X)$, there exist two unital homomorphisms H_1, $H_2 : C(X) \to C([0, 1], A)$ such that

$$\pi_0 \circ H_1 \approx_{\epsilon/3} h_1, \ \ \pi_1 \circ H_1 \approx_{\epsilon/3} \pi_0 \circ H_2 \ \ and \ \ \pi_1 \circ H_2 \approx_{\epsilon/3} h_2 \ \ on \ \mathcal{F}.$$

Moreover,

$$\overline{\text{Length}}(\{\pi_t \circ H_1\}) \le L(X, \xi_1)(1 + 2\pi) + \epsilon/2 \ \ and$$
$$\overline{\text{Length}}(\{\pi_t \circ H_2\}) \le L(X, \xi_2) + \epsilon/2,$$

where ξ_i is (any) point in the spectrum of h_i, $i = 1, 2$.
(Note $L(X, \xi_i) \le \bar{L}_p(X)$).

PROOF. Let $\epsilon > 0$ and $\mathcal{F} \subset C(X)$ be finite subset. Let $\delta > 0$ and $\mathcal{G} \subset C(X)$ be a finite subset required in 10.3 for $\epsilon/3$ and $\mathcal{F}$.

Suppose that ξ_i is a point in the spectrum of h_i, $i = 1, 2$. It follows from virtue of 10.8 and 10.9 that we may write that

$$\|h_1(f) - (h_1'(f) + f(\xi_1)p_1)\| < \epsilon/3 \ \text{and} \tag{e 13.9}$$
$$\|h_2(f) - (h_2'(f) + f(\xi_2)p_2)\| < \epsilon/3 \tag{e 13.10}$$

for all $f \in \mathcal{G}$, where p_1, p_2 are projections in A with $[p_1] = [p_2] = 0$ and $h_i' : C(X) \to (1 - p_i)A(1 - p_i)$ is a unital monomorphism with $[h_i'] = [h_i] = [h_1]$ in $KL(C(X), A)$, $i = 1, 2,$. Denote $\phi_i(f) = f(\xi_i)p_i$ for all $f \in C(X)$. Suppose that $\{x_1, x_2, ..., x_m\} \subset C(X)$ such that

$$\max\{\|f(x_i)\| : i = 1, 2, ..., m\} \ge (3/4)\|f\| \ \text{for all} \ f \in \mathcal{G}. \tag{e 13.11}$$

There are m non-zero mutually orthogonal projections $p_{i,1}, p_{i,2}, ..., p_{i,m} \in p_iAp_i$ such that $[p_{i,1}] = 0$, $i = 1, 2$. Define

$$\phi_i'(f) = \sum_{k=1}^{m} f(x_i)p_{i,k} \ \text{for all} \ f \in C(X), \ \ i = 1, 2. \tag{e 13.12}$$

By 12.11 there is a unital homomorphism $H_i' : C(X) \to C([0, 1], p_iAp_i)$ such that

$$\pi_0 \circ H_i' = h_i' + \phi_i \ \text{and} \ \pi_1 \circ H_i' = h_i' + \phi_i' \tag{e 13.13}$$

$i = 1, 2$. Moreover,

$$\overline{\text{Length}}(\{\pi_t \circ H_i'\}) \le L(X, \xi_i) + \epsilon/2, \ \ i = 1, 2. \tag{e 13.14}$$

Note that both $h_i' + \phi_i$ are $1/2$-$\mathcal{G}$-injective homomorphisms and

$$[h_1' + \phi_1'] = [h_1'] = [h_1] = [h_2' + \phi_2']. \tag{e 13.15}$$

Combining 10.3 and 12.1, we obtain a unitary $u \in A$ with $[u] = 0$ in $K_1(A)$ such that

$$\text{(e 13.16)} \qquad \operatorname{ad} u \circ (h_1' + \phi_1') \approx_{\epsilon/3} h_2' + \phi_2' \text{ on } \mathcal{G}.$$

Let $\{u_t : t \in [0, 1/2]\}$ be a continuous path of unitaries in A such that $u_0 = 1$, $u_1 = u$ and

$$\text{Length}(\{u_t\}) \leq \pi + \epsilon/8(1 + L(X)).$$

Now define $H_1 : C(X) \to C([0,1], A)$ as follows

$$\pi_t \circ H_1 = \begin{cases} \pi_{2t} \circ H_1' & \text{for all } t \in [0, 1/2] \\ \operatorname{ad} u_t \circ (h_1' + \phi_1') & \text{for all } t \in (1/2, 1]. \end{cases}$$

It follows from (e 13.14), 12.8 and (e 12.7) that

$$\overline{\text{Length}}(\{\pi_t \circ H_1\}) \leq L(X, \xi_1) + \epsilon/2 + 2\pi \cdot \underline{L}_p(X) \leq L(X, \xi_1)(1 + 2\pi) + \epsilon/2.$$

Define $H_2 : C(X) \to C([0,1], A)$ by $\pi_t \circ H_2 = \pi_{1-t} \circ H_2'$ for $t \in [0, 1]$. Then, by (e 13.14),

$$\overline{\text{Length}}(\{\pi_t \circ H_2\}) \leq L(X, \xi_2) + \epsilon/2.$$

We then check that H_1 and H_2 meet the requirements of the theorem. □

Lemma 13.6. *Let X be a finite CW complex, let $\epsilon > 0$ and let $\mathcal{F} \subset C(X)$ be a finite subset. Let $\eta = \sigma_{X, \mathcal{F}, \epsilon/16}$. Let $\{x_1, x_2, ..., x_m\}$ be an $\eta/2$-dense subset of X for which $O_i \cap O_j = \emptyset$ (if $i \neq j$), where*

$$O_i = \{x \in X : dist(x, x_i) < \eta/2s\}, \quad i = 1, 2, ..., m$$

for some integer $s \geq 1$. Let $0 < \sigma < 1/2s$.

There exists $\delta > 0$, a finite subset $\mathcal{G} \subset C(X)$ and a finite subset $\mathcal{P} \subset \underline{K}(C(X))$ satisfying the following:

Suppose that A is a unital separable simple C^*-algebra with tracial rank zero and suppose that $\psi : C(X) \to A$ is a unital δ-$\mathcal{G}$-multiplicative contractive completely positive linear map such that

$$\text{(e 13.17)} \qquad [\psi]|_{\mathcal{P}} = [h]|_{\mathcal{P}} \text{ and } \mu_{\tau \circ \psi}(O_i) \geq \sigma \cdot \eta, \; i = 1, 2, ..., m$$

for some homomorphism $h : C(X) \to A$ and for all $\tau \in T(A)$. Then, there exists a monomorphism $\phi : C(X) \to A$ such that

$$\text{(e 13.18)} \qquad \psi \approx_\epsilon \phi \text{ on } \mathcal{F}$$

Proof. To simplify the notation, by considering each component, without loss of generality, we may assume X is connected finite CW complex. Let $\epsilon > 0$ and $\mathcal{F} \subset C(X)$ be a finite subset. Let $\eta > 0$, $\{x_1, x_2, .., x_m\}$, $\sigma > 0$, $s > 0$ and O_i be as in the statement.

Let $\sigma_1 = \sigma/2$. Let $\gamma > 0$, $\mathcal{G}_1 \subset C(X)$, $\delta_1 > 0$ (in place of δ) and $\mathcal{P} \subset \underline{K}(C(X))$ be as required by Theorem 4.6 of [**33**] for $\epsilon/4$ (instead of ϵ), $\mathcal{F}$, σ_1 (in place of σ) and η above.

We may also assume that

$$[L_1]|_{\mathcal{P}} = [L_2]|_{\mathcal{P}}$$

for any pair of δ_1-$\mathcal{G}_1$-multiplicative contractive completely positive linear maps from $C(X)$ to (any) unital C^*-algebra B, provided that

$$L_1 \approx_{\delta_1} L_2 \text{ on } \mathcal{G}_1.$$

Let $\delta_2 > 0$ and $\mathcal{G} \subset C(X)$ be a finite subset required in 5.9 for $\min\{\epsilon/4, \delta_1/2\}$ (in pace of ϵ) and $\mathcal{G}_1$ (in place of $\mathcal{F}$). We may assume that $\delta_2 < \min\{\delta_1, \epsilon/4\}$ and $\mathcal{G}_1 \cup \mathcal{F} \subset \mathcal{G}$.

Let $\delta = \delta_2/4$. Let ψ be as in the statement. Without loss of generality, by applying Lemma 5.9 (with $\psi = \phi$) we may write that

$$\psi(f) \approx_{\delta_1/2} \psi_1 \oplus h_0(f) \text{ for all } f \in \mathcal{G}_1, \tag{e 13.19}$$

where $\psi_1 : C(X) \to (1-p)A(1-p)$ is a unital $\delta_1/2$-$\mathcal{G}_1$-multiplicative contractive completely positive linear map and $h_0 : C(X) \to pAp$ is a homomorphism with finite dimensional range, p is a projection with $\tau(1-p) < \gamma/2$ for all $\tau \in T(A)$. We may further assume that

$$\mu_{\tau\circ h_0}(O_i) > \sigma_1 \cdot \eta/2 = \sigma \cdot \eta, \;\; i = 1, 2, ..., m. \tag{e 13.20}$$

Note that we assume that X is connected. Fix $\xi_X \in X$ and let $Y_X = X \setminus \{\xi_X\}$. Since h_0 is a homomorphism with finite dimensional range,

$$[h_0|_{C_0(Y_X)}] = 0 \text{ in } KL(C_0(Y_X), A). \tag{e 13.21}$$

It follows from 4.2 that there is a unital monomorphism $h_1 : C(X) \to (1-p)A(1-p)$ such that

$$[h_1|_{C_0(Y_X)}] = [h|_{C_0(Y_X)}]. \tag{e 13.22}$$

By (e 13.19), (e 13.22) and (e 13.21),

$$[h_1]|_{\mathcal{P}} = [\psi_1]|_{\mathcal{P}} \text{ in } KK(C(X), A). \tag{e 13.23}$$

By applying Theorem 4.6 of [**33**], we obtain a unitary $u \in A$ such that

$$\mathrm{ad}\, u \circ (h_1 \oplus h_0) \approx_{\epsilon/2} \psi_1 \oplus h_0 \text{ on } \mathcal{F}. \tag{e 13.24}$$

Put $h = \mathrm{ad}\, u \circ (h_1 \oplus h_0)$. Then h is a monomorphism. We have

$$h \approx_{\epsilon} \psi \text{ on } \mathcal{F}. \tag{e 13.25}$$

□

Lemma 13.7. *Let X be a finite CW complex and let A be a unital separable simple C^*-algebra with tracial rank zero. Suppose that $h : C(X) \to A$ is a unital homomorphism with the spectrum F. Then, for any $\epsilon > 0$ and any finite subset $\mathcal{F} \subset C(X)$ and $\gamma > 0$, there is a projection $p \in A$ with $\tau(1-p) < \gamma$ for all $\tau \in T(A)$ and a unital homomorphism $h_0 : C(Y) \to pAp$ with finite dimensional range and a unital monomorphism $\phi : C(Y) \to (1-p)A(1-p)$ for a compact subset Y which contains F and which is a finite CW complex such that*

$$h \approx_{\epsilon} h_0 \circ s_1 \oplus \phi \circ s_1 \text{ on } \mathcal{F},$$

where $s_1 : C(X) \to C(Y)$ is the quotient map.

Proof. Let $\epsilon > 0$, $\gamma > 0$ and let $\mathcal{F} \subset C(X)$ be a finite subset. Let $\eta > 0$ be such that $|f(x) - f(x')| < \epsilon/32$ whenever $\mathrm{dist}(x, x') < 2\eta$ and for all $f \in \mathcal{F}$. In other words, $\eta < \frac{\sigma_{X,\mathcal{F},\epsilon/32}}{2}$.

There is a finite CW complex Y such that $F \subset Y \subset X$ and

$$Y \subset \{x \in X : \mathrm{dist}(x, F) \le \eta/4\}. \tag{e 13.26}$$

Denote by s_1 the quotient map from $C(X)$ onto $C(Y)$. Suppose that $Y = \sqcup_{i=1}^K Y_i$, where each Y_i is a connected component of Y. Let $E_1, E_2, .., E_K$ be mutually orthogonal projections in $C(Y)$ corresponding to the components $Y_1, Y_2, ..., Y_K$.

Define $h_Y : C(Y) \to A$ such that $h_Y \circ s_1 = h$. Let $\{x_1, x_2, ..., x_m\}$ be an $\eta/4$-dense subset of Y. By replacing $x_1, x_2, ..., x_m$ by points in F, we can still assume that they are $\eta/2$-dense in Y. There is an integer $s \geq 1$ for which $O_i \cap O_j = \emptyset$ $(i \neq j)$, where

$$O_i = \{x \in Y : \dim(x, x_i) < \eta/2s\}.$$

Note that $2\eta \leq \sigma_{Y, s_1(\mathcal{F}, \epsilon/32)}$.

Put

$$\sigma_0 = \inf\{\mu_{\tau \circ h_Y}(O_i) : 1 \leq i \leq m, \ \ \tau \in T(A)\}/\eta.$$

Since A is simple and $x_i \in F$, $i = 1, 2, ..., m$, $\sigma_0 > 0$. We may assume that $m > 2$. Choose $\sigma > 0$ so small that it is smaller than $(1 - 1/m)\sigma_0$.

Let $\delta > 0$, $\mathcal{G}$, $\mathcal{P}$ and $\gamma_1 > 0$ be as required by 4.6 of [**33**] for Y corresponding to the above $\epsilon/4$, $s_1(\mathcal{F})$ and σ. Let

$$\gamma_2 = \min\{\gamma, \gamma_1, \sigma_0\eta/m\}.$$

We may assume that

$$[L_1]|_{\mathcal{P}} = [L_2]|_{\mathcal{P}}$$

for any pair of δ-$\mathcal{G}$-multiplicative contractive completely positive linear map $L_1, L_2 : C(Y) \to A$ provided that

$$L_1 \approx_\delta L_2 \text{ on } \mathcal{G}.$$

We may also assume that $E_1, E_2, .., E_K \subset \mathcal{G}$.

Let $\mathcal{G}_1 \subset C(Y)$ be a finite subset such that $\mathcal{G} \subset \mathcal{G}_1$. By applying 5.3, we may write

$$h_Y \approx_{\delta/2} \psi \oplus h_0' \text{ on } \mathcal{G}_1, \tag{e 13.27}$$

where $\psi : C(Y) \to (1-p)A(1-p)$ is a unital δ-$\mathcal{G}_1$-multiplicative contractive completely positive linear map and $h_0' : C(Y) \to pAp$ is a unital homomorphism with finite dimensional range and $p \in A$ is projection in A with $\tau(1-p) < \gamma_2$ for all $\tau \in T(A)$.

With sufficiently large $\mathcal{G}_1$ and smaller δ, we may assume that

$$\mu_{\tau \circ h_0'}(O_i) \geq (1 - 1/m)\sigma_0\eta \geq \sigma\eta$$

for all $\tau \in T(A)$, $i = 1, 2, ..., m$.

By (e 13.27), we may also assume that

$$[h_Y]|_{\mathcal{P}} = [\psi \oplus h_0']|_{\mathcal{P}}. \tag{e 13.28}$$

Suppose that $h_Y(E_i) = e_i$ and $h_0'(E_i) = e_i'$, $i = 1, 2, ..., K$.

Fix a point $\xi_i \in Y_i$ and let $\Omega_i = Y_i \setminus \{\xi_i\}$. By 4.2, there is a unital monomorphism $h_{00}^{(i)} : C(Y_i) \to (e_i - e_i')A(e_i - e_i')$ such that

$$[h_{00}^{(i)}|_{C_0(\Omega_i)}] = [h_Y|_{C_0(\Omega_i)}]. \tag{e 13.29}$$

Define $h_{00} : C(Y) \to (1-p)A(1-p)$ by $h_{00}(f) = \sum_{i=1}^K h_{00}^{(i)}(f|_{Y_i})$ for $f \in C(Y)$. It follows from (e 13.28) that

$$[h_{00}]|_{\mathcal{P}} = [\psi]|_{\mathcal{P}}.$$

It follows from 4.6 of [**33**] that there is a unitary $w \in A$ such that

$$h_0' \oplus \psi \approx_{\epsilon/4} \operatorname{ad} w \circ (h_0' \oplus h_{00}) \text{ on } s_1(\mathcal{F}).$$

Define $h_0 = \operatorname{ad} w \circ h_0'$ and $\phi = \operatorname{ad} w \circ h_{00}$.

From the above, h_0 and ϕ satisfy the requirements. □

Theorem 13.8. *Let X be a connected finite CW complex and let A be a unital separable simple C^*-algebra with tracial rank zero. Suppose that $\psi_1, \psi_2 : C(X) \to A$ are two unital monomorphisms such that*

$$[\psi_1] = [\psi_2] \text{ in } KL(C(X), A). \tag{e 13.30}$$

Then, for any $\epsilon > 0$ and any finite subset $\mathcal{F} \subset C(X)$, there exists a unital homomorphism $H : C(X) \to C([0,1], A)$ such that

$$\pi_0 \circ H \approx_\epsilon \psi_1 \text{ and } \pi_1 \circ H \approx_\epsilon \psi_2 \text{ on } \mathcal{F}. \tag{e 13.31}$$

Moreover, each $\pi_t \circ H$ is a monomorphism and

$$\overline{\text{Length}}(\{\pi_t \circ H\}) \le L(X) + \underline{L}_p(X)2\pi + \epsilon. \tag{e 13.32}$$

PROOF. Let $\delta > 0$ and $\mathcal{G} \subset C(X)$ be a finite subset required by Theorem 3.3 of [**33**] (for ψ_2 being the given unital monomorphism) and for $\epsilon/4$ and $\mathcal{F}$ given. By 13.7 we may write that

$$\psi_1 \approx_{\min\{\epsilon/4, \delta/2\}} h_1 \oplus h_0 \text{ on } \mathcal{G}, \tag{e 13.33}$$

where $h_1 : C(X) \to (1-p)A(1-p)$ is a unital monomorphism, $h_0 : C(X) \to pAp$ is a unital homomorphism with finite dimensional range and $p \in A$ is a projection such that $\tau(1-p) < \delta/2$ for all $\tau \in T(A)$. By 4.3, there is a unital homomorphism $h_{00} : C(X) \to pAp$ with finite dimensional range such that

$$\sup\{|\tau \circ h_{00}(f) - \tau \circ \psi_2(f)| : f \in \mathcal{G}, \tau \in T(A)\} < \delta/2. \tag{e 13.34}$$

Write

$$h_0(f) = \sum_{i=1}^m f(x_i)p_i \text{ and } h_{00}(f) = \sum_{j=1}^m f(x_j)e_j \text{ for all } f \in C(X), \tag{e 13.35}$$

where $x_i, y_j \in X$ and $\{p_1, p_2, ..., p_n\}$ and $\{e_1, e_2, ..., e_m\}$ are two sets of mutually orthogonal projections with $p = \sum_{i=1}^n p_i = \sum_{j=1}^m e_j$. It follows from 5.8 that there is a unital commutative finite dimensional C^*-subalgebra B of pAp which contains $p_1, p_2, ..., p_n$ and a set of mutually orthogonal projections $\{e_1', e_2', ..., e_m'\}$ such that $[e_j'] = [e_j]$, $j = 1, 2, ..., m$. Define

$$h_{00}'(f) = \sum_{j=1}^m f(x_j)e_j' \text{ for all } f \in C(X). \tag{e 13.36}$$

Note, by 12.11, there is a unital homomorphism $\Phi_1 : C(X) \to C([0,1], B)$ such that

$$\pi_0 \circ \Phi_1 = h_0 \text{ for all } f \in C(X), \quad \pi_1 \circ \Phi_1 = h_{00}' \text{ and}$$
$$\overline{\text{Length}}(\{\pi_t \circ \Phi_1\}) \le L(X) + \epsilon/2.$$

It follows that there is a homomorphism $H_1 : C(X) \to pAp$ such that

$$\pi_0 \circ H_1 = h_1 \oplus h_0 \text{ on } \mathcal{F}, \quad \pi_1 \circ H_1 = h_1 \oplus h_{00}' \text{ and} \tag{e 13.37}$$

$$\overline{\text{Length}}(\{\pi_t \circ H_1\}) \le L(X) + \epsilon/2. \tag{e 13.38}$$

Since $[e_j'] = [e_j]$, $j = 1, 2, ..., m$, by (e 13.34),

$$\sup\{|\tau \circ (h_1 \oplus h_{00}')(f) - \tau \circ \psi_2(f)| : \tau \in T(A)\} < \delta$$

for all $f \in \mathcal{G}$.

By applying Theorem 3.3 of [**33**], we obtain a unitary $u \in A$ such that

$$\text{(e 13.39)} \qquad \operatorname{ad} u \circ (h_1 \oplus h_{00}) \approx_{\epsilon/4} \psi_2 \text{ on } \mathcal{F}.$$

It follows from 12.1 that we may assume that $u \in U_0(A)$. It follows that there is a continuous path of unitaries $\{u_t : t \in [0,1]\}$ of A for which

$$\text{(e 13.40)} \quad u_0 = 1_A, \ u_1 = u \text{ and } \operatorname{Length}(\{u_t\}) \le \pi + \frac{\epsilon}{4(1+L(X))}.$$

Connecting $\pi_t \circ H_1$ with $\operatorname{ad} u_t \circ (h_1 \oplus h_{00})$, we obtain a homomorphism $H : C(X) \to C([0,1], A)$ such that

$$\text{(e 13.41)} \qquad \pi_0 \circ H = h_1 \oplus h_0 \text{ on } \mathcal{F}, \ \ \pi_1 \circ H = \operatorname{ad} u \circ (h_1 \oplus h_{00}) \text{ and}$$

$$\text{(e 13.42)} \qquad \overline{\operatorname{Length}}(\{\pi_t \circ H\}) \le L(X) + \underline{L}_p(X)2\pi + \epsilon.$$

It is clear that H meets the requirements. □

Theorem 13.9. *Let X be a connected finite CW complex and let A be a unital separable simple C^*-algebra with tracial rank zero. Suppose that $\psi_1, \psi_2 : C(X) \to A$ are unital homomorphisms such that*

$$[\psi_1] = [\psi_2] \text{ in } KL(C(X), A).$$

Then, for any $\epsilon > 0$ and any finite subset $\mathcal{F} \subset C(X)$, there exist two homomorphisms $H_i : C(X) \to C([0,1], A)$ $(i = 1, 2)$ such that

$$\pi_0 \circ H_1 \approx_{\epsilon/3} \psi_1, \pi_1 \circ H_1 \approx_{\epsilon/3} \pi_0 \circ H_2 \text{ and } \pi_1 \circ H_2 \approx_{\epsilon/3} \psi_2 \text{ on } \mathcal{F}.$$

Moreover,

$$\overline{\operatorname{Length}}(\{\pi_t \circ H_1\}) \le L(X) + 2\pi \underline{L}_p(X) + \epsilon/2 \text{ and}$$
$$\overline{\operatorname{Length}}(\{\pi_t \circ H_2\}) \le L(X) + \epsilon/2.$$

PROOF. It follows from 4.2 that there is a unital monomorphism $\phi : C(X) \to A$ such that

$$\text{(e 13.43)} \qquad [\phi] = [\psi_1] = [\psi_2] \text{ in } KL(C(X), A).$$

Let $\delta > 0$ and $\mathcal{G} \subset C(X)$ be a finite subset required by Theorem 3.3 of [**33**] (for ϕ being the given unital monomorphism) and for $\epsilon/8$ and $\mathcal{F}$ given. By 13.7 we may write that

$$\text{(e 13.44)} \qquad \psi_i \approx_{\min\{\epsilon/4, \delta/2\}} h_i \oplus h_0^{(i)} \text{ on } \mathcal{G}.$$

where $h_i : C(X) \to (1-p_i)A(1-p_i)$ is a unital homomorphism, $h_0^{(i)} : C(X) \to p_iAp_i$ is a unital homomorphism with finite dimensional range and $p_i \in A$ is a projection such that $\tau(1-p_i) < \delta/2$ for all $\tau \in T(A)$, $i = 1, 2$. By 4.3, there is a unital homomorphism $h_{00}^{(i)} : C(X) \to p_iAp_i$ with finite dimensional range such that

$$\text{(e 13.45)} \qquad \sup\{|\tau \circ h_{00}^{(i)}(f) - \tau \circ \phi(f)| : f \in \mathcal{G}, \tau \in T(A)\} < \delta/2.$$

By applying 5.8, 12.11 and the proof of 13.8, without loss of generality, we may assume that there is a unital homomorphism $\Phi_i : C(X) \to C([0,1], p_iAp_i)$ such that

$$\pi_0 \circ \Phi_i(f) = h_0^{(i)}, \ \ \pi_1 \circ \Phi_i = h_{00}^{(i)} \text{ and}$$
$$\overline{\operatorname{Length}}(\{\pi_t \circ \Phi_i\}) \le L(X) + \epsilon/4.$$

It follows that there is a unital homomorphism $H_i': C(X) \to C([0,1], A)$ such that

$$\pi_0 \circ H_i' = h_i \oplus h_0^{(i)}, \quad \pi_1 \circ H_i' = h_i \oplus h_{00}^{(i)} \text{ and} \tag{e 13.46}$$

$$\overline{\text{Length}}(\{\pi_t \circ H_i'\}) \le L(X) + \epsilon/4. \tag{e 13.47}$$

Note that

$$\sup\{|\tau \circ (h_i \oplus h_{00}^{(i)})(f) - \tau \circ \phi(f)| : \tau \in T(A)\} < \delta$$

for all $f \in \mathcal{G}$. By applying Theorem 3.3 of [**33**], we obtain a unitary $u \in A$ such that

$$\text{ad}\, u \circ (h_1 \oplus h_{00}^{(1)}) \approx_{\epsilon/4} h_2 \oplus h_{00}^{(2)} \text{ on } \mathcal{F}. \tag{e 13.48}$$

It follows from 12.1 that we may assume that $u \in U_0(A)$. It follows that there is a continuous path of unitaries $\{u_t : t \in [0,1]\}$ of A for which

$$u_0 = 1_A, \quad u_1 = u \text{ and } \text{Length}(\{u_t\}) \le \pi + \frac{\epsilon}{8(1+L(X))}. \tag{e 13.49}$$

Connecting $\pi_t \circ H_1'$ with $\text{ad}\, u_t \circ (h_1 \oplus h_{00}^{(1)})$, we obtain a homomorphism $H_1 : C(X) \to C([0,1], A)$ such that

$$\pi_0 \circ H_1 = h_1 \oplus h_0^{(1)}, \quad \pi_1 \circ H_1 \approx_{\epsilon/4} h_2 \oplus h_{00}^{(2)} \text{ on } \mathcal{F} \text{ and} \tag{e 13.50}$$

$$\overline{\text{Length}}(\{\pi_t \circ H_1\}) \le L(X) + \underline{L}_p(X) 2\pi + \epsilon. \tag{e 13.51}$$

Define H_2 by $\pi_t \circ H_2' = \pi_{1-t} \circ H_2'$. It is clear that H_1 and H_2 meet the requirements. □

14. Approximate homotopy for approximately multiplicative maps

Theorem 14.1. *Let X be a compact metric space, let $\epsilon > 0$ and let $\mathcal{F} \subset C(X)$ be a finite subset. Then there exits $\delta > 0$, a finite subset $\mathcal{G} \subset C(X)$ and a finite subset $\mathcal{P} \subset \underline{K}(C(X))$ satisfying the following:*

Suppose that A is a unital separable purely infinite simple C^*-algebra and suppose that $\psi_1, \psi_2 : C(X) \to A$ are two δ-$\mathcal{G}$-multiplicative and $1/2$-$\mathcal{G}$-injective contractive completely positive linear maps. If

$$[\psi_1]|_{\mathcal{P}} = [\psi_2]|_{\mathcal{P}}, \tag{e 14.1}$$

then there exists an ϵ-$\mathcal{F}$-multiplicative contractive completely positive linear map $H_1 : C(X) \to C([0,1], A)$ such that

$$\pi_0 \circ H_1 \approx_\epsilon \psi_1, \quad \pi_1 \circ H_1 \approx_\epsilon \psi_2 \text{ on } \mathcal{F}.$$

Moreover,

$$\text{Length}(\{\pi_t \circ H\}) \le 2\pi + \epsilon \tag{e 14.2}$$

$$\overline{\text{Length}}(\{\pi_t \circ H\}) \le 2\pi \underline{L}_p(X) + \epsilon \tag{e 14.3}$$

Proof. Let $\epsilon > 0$ and let $\mathcal{F} \subset C(X)$ be a finite subset. It follows from 10.3, 2.23 and 10.1 that, for a choice of δ, $\mathcal{G}$ and $\mathcal{P}$ as in the statement of the theorem, there exists a unitary $u \in A$ and there exists a projection $p \in A$ such that

$$\text{ad}\, u \circ \psi_1(f) \approx_{\epsilon/2} \psi_2(f) \text{ and } \psi_2(f) \approx_{\epsilon/2} (1-p)\psi_2(f)(1-p) + f(\xi)p \tag{e 14.4}$$

for $f \in \mathcal{F}$ and for some $\xi \in X$. As in the proof 12.1, one finds a unitary $w \in pAp$ with $[1-p+w] = [u]$ in $K_1(A)$. Thus (by replacing $\epsilon/2$ by ϵ), we may assume that

$u \in U_0(A)$. We then, by a result of N. C. Phillips ([**46**]), obtain a continuous path of unitaries $\{u_t : t \in [0,1]\}$ such that

$$u_0 = 1, \ \ u_1 = u \ \text{and} \ \mathrm{Length}(\{u_t\}) \le \pi + \frac{\epsilon}{4(1+L(X))}.$$

Define $H : C(X) \to C([0,1])$ by $\pi_t \circ H = \mathrm{ad}\, u_t \circ h_1$ for $t \in [0,1]$. □

Theorem 14.2. *Let X be a a connected finite CW complex, let $\epsilon > 0$ and let $\mathcal{F} \subset C(X)$ be a finite subset. Then there exits $\delta > 0$, a finite subset $\mathcal{G} \subset C(X)$ and a finite subset $\mathcal{P} \subset \underline{K}(C(X))$ satisfying the following:*

Suppose that A is a unital separable purely infinite simple C^*-algebra and suppose that $\psi_1, \psi_2 : C(X) \to A$ are two unital δ-$\mathcal{G}$-multiplicative contractive completely positive linear maps. If

$$[\psi_1]|_{\mathcal{P}} = [\psi_2]|_{\mathcal{P}}, \tag{e 14.5}$$

then there exist two unital ϵ-$\mathcal{F}$-multiplicative contractive completely positive linear maps $H_1, H_2 : C(X) \to C([0,1], A)$ such that

$$\pi_0 \circ H_1 \approx_{\epsilon/3} \psi_1, \ \ \pi_1 \circ H_1 \approx_{\epsilon/3} \pi_0 \circ H_2 \ \ \textit{and}$$
$$\pi_1 \circ H_2 \approx_{\epsilon/3} \psi_2 \ \text{on} \ \mathcal{F}.$$

Moreover,

$$\overline{\mathrm{Length}}(\{\pi_t \circ H_1\}) \le \bar{L}_p(X)(1+2\pi) + \epsilon \ \textit{and} \tag{e 14.6}$$

$$\overline{\mathrm{Length}}(\{\pi_t \circ H_2\}) \le \bar{L}_p(X) + \epsilon. \tag{e 14.7}$$

Proof. Fix $\epsilon > 0$ and fix a finite subset $\mathcal{F} \subset C(X)$.

Let $\delta_1 > 0$ and let $\mathcal{G}_1 \subset C(X)$ be a finite subset and let $\mathcal{P} \subset \underline{K}(C(X))$ be a finite subset required by 10.3 for $\epsilon/4$ and $\mathcal{F}$. We may assume that

$$[L_1]|_{\mathcal{P}} = [L_2]|_{\mathcal{P}}$$

for any pair of δ_1-$\mathcal{G}_1$-multiplicative contractive completely positive linear maps $L_1, L_2 : C(X) \to A$, provided that

$$L_1 \approx_{\delta_1} L_2 \ \text{on} \ \mathcal{G}_1.$$

Let $\delta_2 > 0$ and a finite subset $\mathcal{G}_2 \subset C(X)$ required by 10.1 corresponding to the finite subset $\mathcal{G}_1$ and positive number $\min\{\epsilon/6, \delta_1/2\}$.

Choose $\delta = \min\{\delta_1, \delta_2, \epsilon/6\}$ and $\mathcal{G} = \mathcal{G}_1 \cup \mathcal{G}_2 \cup \mathcal{F}$.

Now suppose that ψ_1 and ψ_2 are as in the statement with above δ, $\mathcal{G}$ and $\mathcal{P}$. We may write

$$\psi_i \approx_{\min\{\delta_1/2, \epsilon/6\}} \phi_i \oplus h_i \ \text{on} \ \mathcal{G}_1,$$

where $\phi_i : C(X) \to (1-p_i)A(1-p_i)$ is a $\delta_1/2$-$\mathcal{G}_1$-multiplicative contractive completely positive linear map and $h_i : C(X) \to p_iAp_i$ is defined by $h_i(f) = f(\xi_i)p_i$ for all $f \in C(X)$, where $\xi_i \in X$ is a point and p_i is a non-zero projection, $i = 1, 2$. It is clear that we may assume that $[p_i] = 0$ in $K_0(A)$. There are $x_1, x_2, ..., x_m$ in X such that

$$\max\{\|g(x_i)\| : i = 1, 2, ..., m\} \ge (3/4)\|g\| \ \text{for all} \ g \in \mathcal{G}_1. \tag{e 14.8}$$

There are non-zero mutually orthogonal projections $p_{i,1}, p_{i,2}, ..., p_{i,m}$ in p_iAp_i such that $[p_{i,k}] = 0$, $k = 1, 2, ..., m$, and $p_i = \sum_{k=1}^m p_{i,k}$, $i = 1, 2$.

Define $h_0^{(i)}(f) = \sum_{k=1}^m f(x_k)p_{i,k}$ for $f \in C(X)$. Note that

$$[h_0^{(i)}] = 0 \ \text{in} \ KL(C(X), A).$$

By 12.11, there are homomorphisms $H_i': C(X) \to C([0,1], p_iAp_i)$ such that

$$\pi_0 \circ H_i' = h_i \text{ and } \pi_1 \circ H_i' = h_0^{(i)}, i = 1, 2. \tag{e 14.9}$$

Moreover, we can require that

$$\overline{\text{Length}}(\{\pi_t \circ H_i'\}) \le L(X, \xi_i) + \epsilon/4, \quad i = 1, 2. \tag{e 14.10}$$

It follows from (e 14.8) that $\phi_i \oplus h_0^{(i)}$ is δ_1-$\mathcal{G}_1$-multiplicative and $1/2$-$\mathcal{G}_1$-injective.

Thus, by 10.3 there exists a unitary $u \in A$ such that

$$\operatorname{ad} u \circ (\phi_1 \oplus h_0^{(1)}) \approx_{\epsilon/3} \phi_2 \oplus h_0^{(2)} \text{ on } \mathcal{F}. \tag{e 14.11}$$

It follows from 12.1 that we may assume that $u \in U_0(A)$. Thus we obtain a continuous rectifiable path $\{u_t : t \in [0,1]\}$ of A such that

$$u_0 = 1_A, \quad u_1 = u \text{ and} \tag{e 14.12}$$
$$\text{Length}(\{u_t\}) \le \pi + \frac{\epsilon}{4(1 + L(X))}.$$

Define $H_1 : C(X) \to C([0,1], A)$ by

$$\pi_t \circ H_1 = \begin{cases} \phi_1 \oplus \pi_{2t} \circ H_1' & \text{if } t \in [0, 1/2]; \\ \operatorname{ad} u_{2(t-1/2)} \circ (\phi_1 \oplus h_0^{(1)})\}) & \text{if } t \in (1/2, 1]. \end{cases}$$

Then

$$\pi_0 \circ H_1 = \phi_1 \oplus h_1, \quad \pi_1 \circ H_1 \approx_{\epsilon/3} \phi_2 \oplus h_0^{(2)} \text{ on } \mathcal{F}. \tag{e 14.13}$$

Moreover,

$$\overline{\text{Length}}(\{\pi_t \circ H_1\}) \le L(X, \xi_1)(1 + 2\pi) + \epsilon. \tag{e 14.14}$$

We then define $H_2 : C(X) \to C([0,1], A)$ by $\pi_t \circ H_2 = \pi_{1-t} \circ H_2''$. We see that H_1 and H_2 meet the requirements of the theorem. □

Theorem 14.3. *Let X be a metric space which is a connected finite CW complex, let $\epsilon > 0$ and let $\mathcal{F} \subset C(X)$ be a finite subset. Then there exists $\delta > 0$, a finite subset $\mathcal{G} \subset C(X)$ and a finite subset $\mathcal{P} \subset \underline{K}(C(X))$ satisfying the following:*

Suppose that A is a unital separable simple C^*-algebra with tracial rank zero and suppose $\psi_1, \psi_2 : C(X) \to A$ are two unital δ-$\mathcal{G}$-multiplicative contractive completely positive linear maps. If

$$[\psi_1]|_{\mathcal{P}} = [\psi_2]|_{\mathcal{P}}, \tag{e 14.15}$$

then there exist two ϵ-$\mathcal{F}$-multiplicative contractive completely positive linear maps $H_1, H_2 : C(X) \to C([0,1], A)$ such that

$$\pi_0 \circ H_1 \approx_{\epsilon/3} \psi_1, \pi_1 \circ H_1 \approx_{\epsilon/3} \pi_1 \circ H_2, \text{ and} \tag{e 14.16}$$
$$\pi_0 \circ H_2 \approx_{\epsilon/3} \psi_2 \text{ on } \mathcal{F}. \tag{e 14.17}$$

Moreover,

$$\overline{\text{Length}}(\{\pi_t \circ H_1\}) \le L(X) + 2\pi \bar{L}_p(X) + \epsilon \text{ and} \tag{e 14.18}$$
$$\overline{\text{Length}}(\{\pi_t \circ H_2\}) \le L(X) + \epsilon. \tag{e 14.19}$$

PROOF. Fix $\epsilon > 0$ and fix a finite subset $\mathcal{F} \subset C(X)$. Let $\eta = \sigma_{X,\mathcal{F},\epsilon/32}$. Suppose that $\{x_1, x_2, ..., x_m\}$ is $\eta/2$-dense in X. There is $s \geq 1$ such that $O_i \cap O_j = \emptyset$, where

$$O_i = \{x \in X : \text{dist}(x, x_i) < \eta/2s\}, \quad i = 1, 2, ..., m.$$

Put $\sigma = \frac{1}{(2sm+1)}$.

Let δ_1 (in place of δ), $\gamma > 0$, $\mathcal{G}_1$ (in place of $\mathcal{G}$) and $\mathcal{P} \subset \underline{K}(C(X))$ be as required by Theorem 4.6 of [**33**] for $\epsilon/6$ (in place of ϵ) and $\mathcal{F}$ and σ. We also assume that $2m\sigma\eta < 1 - \gamma$. We may assume that $\delta < \epsilon$ and $\mathcal{F} \subset \mathcal{G}$. Furthermore, we assume that

$$[L_1]|_{\mathcal{P}} = [L_2]|_{\mathcal{P}}$$

for any pair of δ_1-$\mathcal{G}_2$-multiplicative contractive completely positive linear maps $L_1, L_2 : C(X) \to A$, provided that

$$L_1 \approx_{\delta_1} L_2 \text{ on } \mathcal{G}_1.$$

Let δ and $\mathcal{G}$ be as in 5.9 corresponding to $\min\{\delta_1/2, \epsilon/3\}$ (instead of ϵ) and $\mathcal{G}_1$ (instead of $\mathcal{G}$) and $\gamma/2$.

Let A be a unital separable simple C^*-algebra with tracial rank zero and $\psi_1, \psi_2 : C(X) \to A$ be as described. By 5.9, we may write

$$\psi_i \approx_{\min\{\delta_1/2, \epsilon/6\}} \phi_i \oplus h_i \text{ on } \mathcal{G}_1 \ (i = 1, 2), \tag{e 14.20}$$

where $\phi_i : C(X) \to (1-p_i)A(1-p_i)$ is a unital contractive completely positive linear map and $h_i : C(X) \to p_iAp_i$ is a unital homomorphism with finite dimensional range.

Note that we assume that p_1 and p_2 are unitarily equivalent and

$$\tau(1 - p_i) < \gamma/2 \text{ for all } \tau \in T(A), \ i = 1, 2. \tag{e 14.21}$$

There are non-zero mutually orthogonal projections $p_{i,1}, p_{i,2}, ..., p_{i,m} \in p_iAp_i$ such that $\sum_{k=1}^m p_{i,k} = p_i$ and

$$\tau(p_{i,k}) \geq \frac{1 - \gamma/2}{m + 1}, \quad k = 1, 2, ..., m \text{ and } i = 1, 2. \tag{e 14.22}$$

Since $[p_1] = [p_2]$, we also require that

$$[p_{1,k}] = [p_{2,k}], \quad k = 1, 2, ..., m. \tag{e 14.23}$$

Define $h_0^{(i)} : C(X) \to p_iAp_i$ by

$$h_0^{(i)}(f) = \sum_{k=1}^m f(x_k)p_{i,k} \text{ for all } f \in C(X), \ i = 1, 2.$$

By 5.8, without loss of generality, we may assume that there is a unital commutative finite dimensional C^*-subalgebra $B_i \subset p_iAp_i$ such that $h_i(C(X)), h_0^{(i)}(C(X)) \subset B_i$, $i = 1, 2$.

It follows from 12.11 that there is a homomorphism $H_i' : C(X) \to p_iAp_i$ such that

$$\pi_0 \circ H_i' = h_i, \quad \pi_1 \circ H_i' = h_0^{(i)} \text{ and} \tag{e 14.24}$$

$$\overline{\text{Length}}(\{\pi_t \circ H_i'\}) \leq L(X) + \epsilon/4, \quad i = 1, 2. \tag{e 14.25}$$

Note that, since X is path connected,

$$[h_1] = [h_0^{(1)}] = [h_0^{(2)}] = [h_2] \text{ in } KL(C(X), A). \tag{e 14.26}$$

It follows from (e 14.15), (e 14.20) as well as (e 14.26) that

$$[\psi_1 \oplus h_0^{(1)}]|_{\mathcal{P}} = [\psi_2 \oplus h_0^{(2)}]|_{\mathcal{P}}. \tag{e 14.27}$$

We also have that

$$\mu_{\tau\circ(\psi_i\oplus h_0^{(i)})}(O_k) \geq \tau(p_{i,k}) \geq \sigma\eta \tag{e 14.28}$$

for all $\tau \in T(A)$, $k = 1, 2, ..., m$, $i = 1, 2$. Moreover,

$$|\tau(\phi_1 \oplus h_0^{(1)})(f) - \tau(\phi_2 \oplus h_0^{(2)})(f)| < \gamma \tag{e 14.29}$$

for all $\tau \in T(A)$ and all $f \in \mathcal{G}_1$.

Thus, by Theorem 4.6 of [**33**] there exists a unitary $u \in A$ such that

$$\operatorname{ad} u \circ (\phi_1 \oplus h_0^{(1)}) \approx_{\epsilon/6} \phi_2 \circ h_0^{(2)} \quad \text{on } \mathcal{F}. \tag{e 14.30}$$

By 12.1, we may assume that $u \in U_0(A)$. Thus we obtain a continuous rectifiable path $\{u_t : t \in [0,1]\}$ of A such that

$$u_0 = 1_A, \quad u_1 = u \quad \text{and} \tag{e 14.31}$$

$$\text{Length}(\{u_t\}) \leq \pi + \frac{\epsilon}{4(1 + L(X))}. \tag{e 14.32}$$

Define $H_1 : C(X) \to C([0,1], A)$ by

$$\pi_t \circ H_1 = \begin{cases} \phi_1 \oplus \pi_{2t} \circ H_1' & \text{if } t \in [0, 1/2]; \\ \operatorname{ad} u_{2(t-1/2)} \circ (\phi_1 \oplus h_0^{(1)}) & \text{if } t \in (1/2, 1]. \end{cases}$$

Then

$$\pi_0 \circ H_1 = \phi_1 \oplus h_1, \quad \pi_1 \circ H_1 \approx_{\epsilon/3} \phi_2 \oplus h_0^{(2)} \quad \text{on } \mathcal{F}. \tag{e 14.33}$$

Moreover,

$$\overline{\text{Length}}(\{\pi_t \circ H_1\}) \leq L(X) + 2\pi L(X, \xi_1) + \epsilon. \tag{e 14.34}$$

We then define $H_2 : C(X) \to C([0,1], A)$ by $\pi_t \circ H_2 = \pi_{1-t} \circ H_2''$. We see that H_1 and H_2 meet the requirements of the theorem. □

Remark 14.4. It should be noted that the results in this section are not generalizations of those in the previous section. It is important to know that $\pi_t \circ H_i$ in Theorem 14.2 and those in Theorem 14.3 are not homomorphisms, while $\pi_t \circ H_i$ in 13.9 and 13.5 are unital homomorphisms. Furthermore, $\pi_t \circ H$ in 13.8 and 13.3 are unital monomorphisms.

CHAPTER 5

Super Homotopy

15. Super Homotopy Lemma — the purely infinite case

In this section and the next we study the so-called Super Homotopy Lemma of [**3**] for higher dimensional spaces.

Theorem 15.1. *Let X be a path connected metric space and let $\mathcal{F}\subset C(X)$ be a finite subset.*

Then, for any $\epsilon>0$, there exist $\delta>0$, a finite subset $\mathcal{G}\subset C(X)$ and a finite subset $\mathcal{P}\subset \underline{K}(C(X))$ satisfying the following:

Suppose that A is a unital separable purely infinite simple C^*-algebra and suppose that $h_1, h_2: C(X)\to A$ are two unital monomorphisms such that

$$[h_1]=[h_2] \text{ in } KL(C(X),A). \tag{e 15.1}$$

If there are unitaries $u, v\in A$ such that

$$\|[h_1(f),u]\|<\delta,\ \ \|[h_2(f),v]\|<\delta \ \text{ for all } f\in\mathcal{G} \text{ and} \tag{e 15.2}$$

$$\mathrm{Bott}(h_1,u)|_{\mathcal{P}}=\mathrm{Bott}(h_2,v)|_{\mathcal{P}}, \tag{e 15.3}$$

then there exists a unital monomorphism $H: C(X)\to C([0,1],A)$ and there exists a continuous rectifiable path $\{u_t: t\in[0,1]\}$ in A such that, for each $t\in[0,1]$, $\pi_t\circ H$ is a unital monomorphism,

$$u_0=u,\ \ u_1=v,\ \ \pi_0\circ H\approx_{\epsilon} h_1 \text{ and } \pi_1\circ H\approx_{\epsilon} h_2 \text{ on } \mathcal{F} \tag{e 15.4}$$

and

$$\|[u_t, \pi_t\circ H(f)]\|<\epsilon \tag{e 15.5}$$

for all $t\in[0,1]$ and for all $f\in\mathcal{F}$. Moreover,

$$\mathrm{Length}(\{\pi_t\circ H\}) \ \le\ 2\pi+\frac{\epsilon}{8(1+L(X))}, \tag{e 15.6}$$

$$\overline{\mathrm{Length}}(\{\pi_t\circ H\}) \ \le\ 2\pi\underline{L}_p(X)+\epsilon \text{ and} \tag{e 15.7}$$

$$\mathrm{Length}(\{u_t\}) \ \le\ 4\pi+\epsilon. \tag{e 15.8}$$

(See 12.7 and 12.5 for the definitions of $\overline{\mathrm{Length}}$ and $\underline{L}_p(X)$).

Proof. Let $\epsilon>0$ and let $\mathcal{F}\subset C(X)$ be a finite subset. We may assume that $1_{C(X)}\in\mathcal{F}$. Put

$$\epsilon_1=\min\{\epsilon/8, 2\sin(\epsilon/16)\}.$$

Let $\mathcal{G}_0\subset C(X\times S^1)$ be a finite subset, $\mathcal{P}_1\subset \underline{K}(C(X\times S^1))$ be a finite subset, $\delta_1>0$ be a positive number associated with $\epsilon_1/2$, and $\mathcal{F}\otimes S\subset C(X\times S^1)$ as required by 10.3, where $S=\{1_{C(S^1)}, z\}\subset C(S^1)$. By choosing even smaller δ_1, we may assume that $\mathcal{G}_0=\mathcal{G}_1\otimes S$. Let $\mathcal{P}_2=(\mathrm{id}-\hat{\boldsymbol{\beta}})(\mathcal{P}_1)\cup\hat{\boldsymbol{\beta}}(\mathcal{P}_1)$.

We may also assume that $\delta_1 < \epsilon_1$ and $\mathcal{F} \subset \mathcal{G}_1$. Moreover, we may further assume that

$$[L_1]|_{\mathcal{P}_1 \cup \mathcal{P}_2} = [L_2]|_{\mathcal{P}_1 \cup \mathcal{P}_2}$$

for any pair of δ_1-$\mathcal{G}_0$-multiplicative unital contractive completely positive linear maps $L_1, L_2 : C(X \times S^1) \to A$, provided that

$$L_1 \approx_{\delta_1} L_2 \text{ on } \mathcal{G}_0.$$

Let $\delta_2 > 0$ and let $\mathcal{G}_2$ be a finite subset required by 11.2 corresponding to $\mathcal{G}_1$ (in place of $\mathcal{F}$) and $\delta_1/2$ (in place of ϵ). We may assume that $\delta_2 < \delta_1/2$ and $\mathcal{G}_1 \subset \mathcal{G}_2$.

Let $\delta > 0$ and let $\mathcal{G} \subset C(X)$ (in place of $\mathcal{F}_1$) be a finite subset required in 2.8 for $\delta_2/2$ (in place of ϵ) and $\mathcal{G}_2$ (in place of $\mathcal{F}$).

Let $\mathcal{P} \in \underline{K}(C(X))$ be a finite subset such that $\beta(\mathcal{P}) \supset \hat{\beta}(\mathcal{P}_1)$. Now suppose that h_1, h_2 and u and v are as in the statement of the theorem with the above δ, $\mathcal{G}$ and $\mathcal{P}$.

By 2.8, there are $\delta_2/2$-$\mathcal{G}_2 \otimes S$-multiplicative unital contractive completely positive linear maps $\psi_1, \psi_2 : C(X) \to A$ such that

$$\|\psi_1(f \otimes g) - h_1(f)g(u)\| < \delta_2/2 \quad \text{and} \tag{e 15.9}$$
$$\|\psi_2(f \otimes g) - h_2(f)g(v)\| < \delta_2/2 \tag{e 15.10}$$

for all $f \in \mathcal{G}_2$ and $g \in S$.

By applying 11.2, we obtain two continuous paths of unitaries $\{w_{i,t} : t \in [0,1]\}$ in A and two unital monomorphisms $\Phi_i : C(X \times S^1) \to A$ such that

$$w_{1,0} = u, \;\; w_{2,0} = v, \;\; w_{i,1} = \Phi_i(1 \otimes z), \tag{e 15.11}$$
$$\|\Phi_i(f \otimes 1) - h_i(f)\| < \delta_1/2 \quad \text{and} \quad \|[h_i(f), w_{i,t}]\| < \delta_1/2 \tag{e 15.12}$$

for all $f \in \mathcal{G}_1$ and all $t \in [0,1]$, $i = 1, 2$. Moreover,

$$\text{Length}(\{w_{i,t}\}) \le \pi + \epsilon/8. \tag{e 15.13}$$

It follows from (e 15.1), (e 15.3), (e 15.9), (e 15.10) and 2.7 that

$$[\Phi_1]|_{\mathcal{P}_2} = [\Phi_2]|_{\mathcal{P}_2}. \tag{e 15.14}$$

Since Φ_1 and Φ_2 are homomorphisms, we have

$$[\Phi_1]|_{\mathcal{P}_1} = [\Phi_2]|_{\mathcal{P}_1}. \tag{e 15.15}$$

It follows from 10.3 that there is a unitary $U \in A$ such that

$$\text{ad}\, U \circ \Phi_1 \approx_{\frac{\epsilon_1}{2}} \Phi_2 \;\; \text{for all} \;\; f \in \mathcal{F} \otimes S. \tag{e 15.16}$$

By 12.1, we may assume that $U \in U_0(A)$ (by replacing $\epsilon_1/2$ by ϵ_1). It follows that there is a continuous path of unitaries $\{U_t : t \in [0,1]\}$ such that

$$U_0 = 1, \;\; U_1 = U \;\; \text{and} \;\; \text{Length}(\{U_t\}) \le \pi + \frac{\epsilon}{8(1 + L(X))}. \tag{e 15.17}$$

By (e 15.16), there exists $a \in A_{s.a}$ such that $\|a\| \le \epsilon/8$ such that

$$(\text{ad}\,(U \circ \Phi_1(1 \otimes z))^* \Phi_2(1 \otimes z) = \exp(ia). \tag{e 15.18}$$

Moreover

$$\|\exp(ita) - 1\| < \epsilon/8 \;\; \text{for all} \;\; t \in [0,1]. \tag{e 15.19}$$

Put $z_t = \mathrm{ad}\, U \circ \Phi_1(1 \otimes z)\exp(ita)$. Then

$$z_0 = \mathrm{ad}\, U \circ \Phi_1(1 \otimes z), \;\; z_1 = \Phi_2(1 \otimes z) \text{ and} \tag{e 15.20}$$

$$\mathrm{Length}(\{z_t\}) \le \epsilon/8. \tag{e 15.21}$$

Define u_t as follows

$$u_t = \begin{cases} w_{1,3t} & \text{if } t \in [0, 1/3]; \\ U^*_{3(t-1/3)}\Phi_1(1 \otimes z)U_{3(t-1/3)} & \text{if } t \in (1/3, 2/3]; \\ z_{6(t-2/3)} & \text{if } t \in (2/3, 5/6]; \\ w_{2,3(2/3-t)} & \text{if } t \in (5/6, 1] \end{cases}$$

Define $H : C(X \times S^1) \to A$ by

$$\pi_t \circ H = \begin{cases} h_1 & \text{if } t \in [0, 1/3]; \\ \mathrm{ad}\, U_{3(t-1/3)} \circ h_1 & \text{if } t \in (1/3, 2/3]; \\ \mathrm{ad}\, U_1 \circ h_1 & \text{if } t \in (2/3, 1] \end{cases}$$

We compute that

$$\mathrm{Length}(\{u_t\}) \le 2\pi + 2\epsilon/8 + 2\pi + 2\epsilon/8 + \epsilon/8 < 4\pi + \epsilon \text{ and} \tag{e 15.22}$$

$$\mathrm{Length}(\{\pi_t \circ H\}) \le 2\pi + \frac{\epsilon}{4(1 + L(X, \xi_X))}. \tag{e 15.23}$$

In particular, by 12.8,

$$\overline{\mathrm{Length}}(\{\pi_t \circ H\}) \le 2\pi \underline{L}_p(X) + \epsilon/4. \tag{e 15.24}$$

It follows from (e 15.12) and (e 15.19) that

$$\|[\pi_t \circ H(f), u_t]\| < 2(\delta_1/2) + \delta_1/2 + 2(\epsilon/8) < \epsilon \text{ for all } f \in \mathcal{F}. \tag{e 15.25}$$

We see that $\{u_t : t \in [0, 1]\}$ and H meet the requirements. □

Theorem 15.2. *Let X be a connected metric space and let $\mathcal{F} \subset C(X)$ be a finite subset.*

Then, for any $\epsilon > 0$, there exists $\delta > 0$, a finite subset $\mathcal{G} \subset C(X)$ and a finite subset $\mathcal{P} \subset \underline{K}(C(X))$ satisfying the following:

Suppose that A is a unital separable purely infinite simple C^*-algebra and suppose that $h_1, h_2 : C(X) \to A$ are two unital homomorphisms such that

$$[h_1] = [h_2] \text{ in } KL(C(X), A). \tag{e 15.26}$$

If there are unitaries $u, v \in A$ such that

$$\|[h_1(f), u]\| < \delta, \;\; \|[h_2(f), v]\| < \delta \text{ for all } f \in \mathcal{G} \text{ and} \tag{e 15.27}$$

$$\mathrm{Bott}(h_1, u)|_{\mathcal{P}} = \mathrm{Bott}(h_2, v)|_{\mathcal{P}}, \tag{e 15.28}$$

then there exist two unital homomorphisms $H_1, H_2 : C(X) \to C([0, 1], A)$ and a continuous rectifiable path $\{u_t : t \in [0, 2]\}$ in A such that, for each $t \in [0, 1]$, $\pi_t \circ H$ is a unital homomorphism,

$$u_0 = u, \;\; u_2 = v, \tag{e 15.29}$$

$$\pi_0 \circ H_1 \approx_\epsilon h_1, \;\; \pi_1 \circ H_1 \approx_\epsilon \pi_0 \circ H_2 \text{ and } \pi_1 \circ H_1 \approx_\epsilon h_2 \text{ on } \mathcal{F}, \tag{e 15.30}$$

$$\|[u_t, \pi_t \circ H_1(f)]\| < \epsilon \text{ and} \tag{e 15.31}$$

$$\|[u_{1+t}, \pi_t \circ H_2(f)]\| < \epsilon \text{ for all } t \in [0, 1] \tag{e 15.32}$$

and for all $f \in \mathcal{F}$. Moreover,

$$\overline{\text{Length}}(\{\pi_t \circ H_1 : t \in [0,1/2]\}) \le \bar{L}_p(X) + \epsilon/2,$$

$$\text{Length}(\{\pi_t \circ H_1 : t \in [1/2,1]\}) \le 2\pi + \frac{\epsilon}{8(1+L(X))}, \quad \text{(e 15.33)}$$

$$\overline{\text{Length}}(\{\pi_t \circ H_1\}) \le \bar{L}_p(X) + 2\pi \underline{L}_p(X) + \epsilon, \quad \text{(e 15.34)}$$

$$\overline{\text{Length}}(\{\pi_t \circ H_2\}) \le \bar{L}_p(X) + \epsilon \text{ and} \quad \text{(e 15.35)}$$

$$\text{Length}(\{u_t\}) \le 4\pi + \epsilon. \quad \text{(e 15.36)}$$

PROOF. Let $\epsilon > 0$ and let $\mathcal{F} \subset C(X)$ be a finite subset. We may assume that $1_{C(X)} \in \mathcal{F}$ and $\mathcal{F}$ is in the unit ball of $C(X)$. Put

$$\epsilon_1 = \min\{\epsilon/8, \sin(\epsilon/16)\}.$$

Let $\mathcal{G}_0 \subset C(X \times S^1)$ be a finite subset, $\mathcal{P}_1 \subset \underline{K}(C(X \times S^1))$ be a finite subset, $\delta_1 > 0$ be a positive number associated with $\epsilon_1/2$ and $\mathcal{F} \otimes S \subset C(X \times S^1)$ as required by 10.3, where $S = \{1_{C(S^1)}, z\} \subset C(S^1)$. By choosing even smaller δ_1, we may assume that $\mathcal{G}_0 = \mathcal{G}_1 \otimes S$. Let $\mathcal{P}_2 = (\mathrm{id} - \hat{\beta})(\mathcal{P}_1) \cup \hat{\beta}(\mathcal{P}_1)$.

We may also assume that $\delta_1 < \epsilon_1$ and $\mathcal{F} \subset \mathcal{G}_1$. Moreover, we may further assume that

$$[L_1]|_{\mathcal{P}_1 \cup \mathcal{P}_2} = [L_2]|_{\mathcal{P}_1 \cup \mathcal{P}_2}$$

for any pair of δ_1-$\mathcal{G}_0$-multiplicative unital contractive completely positive linear maps $L_1, L_2 : C(X \otimes S^1) \to A$, provided that

$$L_1 \approx_{\delta_1} L_2 \text{ on } \mathcal{G}_0.$$

Let $\delta_2 > 0$ and let $\mathcal{G}_2 \subset C(X)$ be a finite subset required by 5.5 for $\delta_1/4$ (in place of ϵ) and $\mathcal{G}_0$. We may assume that $\delta_2 < \delta_1/4$ and $\mathcal{G}_0 \subset \mathcal{G}_2$.

Let $\delta > 0$ and let $\mathcal{G} \subset C(X)$ (in place of $\mathcal{F}$) be a finite subset required in 2.8 for $\delta_2/2$ (in place of ϵ) and $\mathcal{G}_2$ (in place of $\mathcal{F}$).

Let $\mathcal{P} \in \underline{K}(C(X))$ be a finite subset such that $\beta(\mathcal{P}) \supset \hat{\beta}(\mathcal{P}_1)$.

Now suppose that h_1, h_2 and u and v are as in the statement of the theorem with the above δ, $\mathcal{G}$ and $\mathcal{P}$. Suppose that the spectrum of h_i is F_i, $i = 1, 2$. Let $s_i : C(X) \to C(F_i)$ be the quotient map, $i = 1, 2$. Define $\bar{h}_i : C(F_i) \to A$ such that $\bar{h}_i \circ s_i = h_i$, $i = 1, 2$.

Let $\eta_i > 0$, let $\mathcal{G}_3^{(i)} \subset C(F_i)$ be a finite subset and let $\mathcal{P}_2^{(i)} \subset \underline{K}(C(F_i))$ be a finite subset required by 10.4 corresponding to $s_i(\mathcal{G}_1)$ (in place of $\mathcal{F}$) and $\delta_1/4$ which work for F_i, $i = 1, 2$. We may assume that $\eta_i < \delta_1/4$ and $s_i(\mathcal{G}_2) \subset \mathcal{G}_3^{(i)}$, $i = 1, 2$.

We may further assume that $[L^{(i)}]|_{\mathcal{P}_2^{(i)}}$ is well defined for any η_i-$\mathcal{G}_3^{(i)}$-multiplicative contractive completely positive linear map $L^{(i)} : C(F_i) \to A$. Put $\eta = \min\{\eta_1, \eta_2\}$ and let $\mathcal{G}_3 \subset C(X)$ be a finite subset such that $s_i(\mathcal{G}_3) \supset \mathcal{G}_3^{(i)}$, $i = 1, 2$.

By 2.8, there are $\delta_2/2$-$\mathcal{G}_2 \otimes S$-multiplicative unital contractive completely positive linear maps $\psi_1, \psi_2 : C(X) \to A$ such that

$$\|\psi_1(f \otimes g) - h_1(f)g(u)\| < \delta_2/2 \quad \text{(e 15.37)}$$

for all $f \in \mathcal{G}_2$ and $g \in S$, and

$$\|\psi_2(f \otimes g) - h_2(f)g(v)\| < \delta_2/2 \quad \text{(e 15.38)}$$

for all $f \in \mathcal{G}_2$ and $g \in S$.

By applying 5.5 (and see the remark 5.6), there is a nonzero projection $p_i \in A$ and $\xi_i \in F_i \subset X$ and $t_i \in S^1$ $(i = 1, 2)$ such that

$$\|h_1(f)g(u) - ((1-p_1)h_1(f)g(u)(1-p_1) + f(\xi_1)g(t_1)p_1)\| < \delta_1/4 \tag{e 15.39}$$

$$\|h_2(f)g(v) - ((1-p_2)h_2(f)g(v)(1-p_2) + f(\xi_2)g(t_1)p_2)\| < \delta_1/4 \tag{e 15.40}$$

for all $f \in \mathcal{G}_1$ and $g \in S$,

$$\|h_i(f) - ((1-p)h_i(f)(1-p) + f(\xi_i)p_i)\| < \eta/2 \text{ for all } f \in \mathcal{G}_3, \tag{e 15.41}$$

$i = 1, 2$. Moreover, we may assume that $(1 - p_i)\bar{h}_i(1 - p_i)$ is η_i-$\mathcal{G}_3^{(i)}$-multiplicative, $i = 1, 2$. Furthermore, there are unitaries $w_i \in (1-p_i)A(1-p_i)$ $(i = 1, 2)$ such that

$$\|u - w_1 \oplus t_1p_1\| < \delta_1/2 \text{ and } \|v - w_2 \oplus t_2p_2\| < \delta_1/2. \tag{e 15.42}$$

(Note that η_i and $\mathcal{G}_3^{(i)}$ can be chosen, which depend on F_i among other things, after δ and $\mathcal{G}$ are chosen, thanks to 5.5.)

By replacing p_i by a non-zero sub-projection, we may assume that $[p_i] = 0$ in $K_0(A)$ (see 5.6) and $(1 - p_i)\bar{h}_i(1 - p_i)$ is $1/2$-$\mathcal{G}_3^{(i)}$-injective.

It follows that

$$[(1-p)\bar{h}_i(1-p)]|_{\mathcal{P}_2^{(i)}} = [\bar{h}_i]|_{\mathcal{P}_2^{(i)}}, \quad i = 1, 2. \tag{e 15.43}$$

It follows from 10.4 that there is a unital homomorphism $h_i' : C(X) \to (1-p_i)A(1-p_i)$ (with spectrum F_i) such that

$$h_i' \approx_{\delta_1/4} (1 - p_i)h_i(1 - p_i) \text{ on } \mathcal{G}_1. \tag{e 15.44}$$

Let $\{x_1, x_2, ..., x_m\} \subset X$ and $\{\zeta_1, \zeta_2, ..., \zeta_k\} \subset S^1$ such that

$$\max\{\|f(x_j \times \zeta_l)\| : 1 \le j \le m, 1 \le l \le k\} \ge (3/4)\|f\| \text{ for all } f \in \mathcal{G}_0. \tag{e 15.45}$$

There are mutually orthogonal non-zero projections $\{e_{j,l}^{(i)} : 1 \le j \le m \text{ and } 1 \le l \le k\}$ in p_iAp_i such that

$$\sum_{j,l} e_{j,l}^{(i)} = p_i \text{ and } [e_{j,l}^{(i)}] = 0 \text{ in } K_0(A),$$

$i = 1, 2$. Define $h_{00}^{(i)} : C(X \times S^1) \to p_iAp_i$ by

$$h_{00}^{(i)}(f) = \sum_{j,l} f(x_j \times \zeta_l)e_{j,l}^{(i)} \text{ for all } f \in C(X \times S^1). \tag{e 15.46}$$

It follows from 12.11 that there is a unital homomorphism $\Phi_i : C(X \times S^1) \to C([0, 1], p_iAp_i)$ such that

$$\pi_0 \circ \Phi_i(f) = f(\xi_i \times t_i)p_i \text{ and } \pi_1 \circ \Phi_i(f) = h_{00}^{(i)}(f) \tag{e 15.47}$$

for all $f \in C(X \times S^1)$ and

$$\overline{\text{Length}}(\{\pi_t \circ \Phi_i|_{C(X)\otimes 1}\}) \le \bar{L}_p(X) + \epsilon/4 \tag{e 15.48}$$

$$\text{Length}(\{\pi_t \circ \Phi_i(1 \otimes z)\}) \le \pi, \tag{e 15.49}$$

$i = 1, 2$.

Now $(1 - p_1)\psi_1(1 - p_1) \oplus h_{00}^{(1)}$ and $(1 - p_2)\psi_2(1 - p_2) \oplus h_{00}^{(2)}$ are both $\delta_1/2$-$\mathcal{G}_1 \otimes S$-multiplicative, by (e 15.37), (e 15.38), (e 15.39) and (e 15.40), and are both $1/2$-$\mathcal{G}_0$-injective, by (e 15.45). Put $L_i' = (1 - p_1)\psi_1(1 - p_1) \oplus h_{00}^{(1)}$, $i = 1, 2$.

Moreover, since $[p_i] = 0$ in $K_0(A)$ and $[h_{00}^{(i)}] = 0$ in $KL(C(X), A)$, by the assumption of δ_1 and $\mathcal{G}_0$, and by the assumption (e 15.26) and (e 15.28),

$$[L_1']|_{\mathcal{P}_1} = [L_2']|_{\mathcal{P}_1}.$$

It follows from 10.3 that there is a unitary $U \in A$ such that

$$\mathrm{ad}\, U \circ L_1' \approx_{\epsilon_1/2} L_2' \text{ on } \mathcal{F} \otimes S. \tag{e 15.50}$$

By 12.1, without loss of generality, we may assume that $U \in U_0(A)$, by replacing $\epsilon_1/2$ by ϵ_1 above. It follows that we obtain a continuous path of unitaries $\{U_t : t \in [0,1]\}$ such that

$$U_0 = 1, \ U_1 = U \text{ and } \mathrm{Length}(\{U_t\}) \le \pi + \frac{\epsilon}{8(1 + L(X))}. \tag{e 15.51}$$

By (e 15.50), there exists $a \in A_{s.a}$ such that $\|a\| \le \epsilon/8$ such that

$$(U^*(w_1 \oplus h_{00}^{(1)}(1 \otimes z))U)^*(w_2 \oplus h_{00}^{(2)}(1 \otimes z)) = \exp(ia). \tag{e 15.52}$$

Moreover

$$\|\exp(ita) - 1\| < \epsilon/8 \text{ for all } t \in [0,1]. \tag{e 15.53}$$

Put $z_t = U^*(w_1 \oplus h_{00}^{(1)}(1 \otimes z))U \exp(ita)$. Then

$$z_0 = U^*(w_1 \oplus (1 \otimes z))U, \ \ z_1 = w_2 \oplus h_{00}^{(2)}(1 \otimes z) \text{ and} \tag{e 15.54}$$

$$\mathrm{Length}(\{z_t\}) \le \epsilon/8. \tag{e 15.55}$$

Similarly, by (e 15.42), we obtain two continuous paths of unitaries $\{V_{1,t} : t \in [0,1]\}$ of A such that

$$V_{1,0} = u, \ V_{2,0} = v, \ V_{i,1} = w_i \oplus t_i p_i, \ \mathrm{Length}(\{V_{i,t}\}) \le \epsilon/8 \text{ and} \tag{e 15.56}$$

$$\|V_{i,t} - 1\| < \epsilon/8. \tag{e 15.57}$$

Define u_t as follows

$$u_t = \begin{cases} V_{1,4t} & \text{if } t \in [0, 1/4]; \\ w_1 \oplus \pi_{4(t-1/4)} \Phi_1(1 \otimes z) & \text{if } t \in (1/4, 1/2]; \\ U^*_{2(t-1/2)}(w_1 \oplus \Phi_1(1 \otimes z))U_{2(t-1/2)} & \text{if } t \in (1/2, 1]; \\ z_{2(t-1)} & \text{if } t \in (1, 3/2]; \\ w_2 \oplus \pi_{4(7/4-t)} \circ \Phi_2(1 \otimes z) & \text{if } t \in (3/2, 7/4]; \\ V_{2,4(2-t)} & \text{if } t \in (7/4, 2]. \end{cases}$$

Define $H_1 : C(X) \to C([0,1], A)$ by

$$\pi_t \circ H_1 = \begin{cases} h_1' \oplus \pi_0 \circ \Phi_1|_{C(X) \otimes 1} & \text{if } t \in [0, 1/4]; \\ h_1' \oplus \pi_{4(t-1/4)} \circ \Phi_1|_{C(X) \otimes 1} & \text{if } t \in (1/4, 1/2]; \\ \mathrm{ad}\, U_{2(t-1/2)} \circ (h_1' \oplus \Phi_1|_{C(X) \otimes 1}) & \text{if } t \in (1/2, 1] \end{cases}$$

and define $H_2 : C(X) \to C([0,1], A)$ by

$$\pi_t \circ H_2 = \begin{cases} h_2' \oplus h_{00}^{(2)} & \text{if } t \in [1, 3/2]; \\ h_2' \oplus \pi_{4(7/4-t)} \circ \Phi_2|_{C(X) \otimes 1} & \text{if } t \in (3/2, 7/4]; \\ h_2' \oplus \pi_0 \circ \Phi_2|_{C(X) \otimes 1} & \text{if } t \in (7/4, 2]. \end{cases}$$

We compute that

$$\text{Length}(\{u_t\}) \le \epsilon/8 + \pi + 2\pi + \epsilon/8 + \pi + \epsilon/8 < 4\pi + \epsilon, \tag{e 15.58}$$

$$\overline{\text{Length}}(\{\pi_t \circ H_1 : t \in [0, 1/2]\}) \le \epsilon/8 + \bar{L}_p(X) + \epsilon/4 \text{ and} \tag{e 15.59}$$

$$\text{Length}(\{\pi_t \circ H_1 : t \in [1/2, 1]\}) \le 2\pi + \frac{\epsilon}{8(1 + L(X))}. \tag{e 15.60}$$

Therefore (by 12.8)

$$\overline{\text{Length}}(\{\pi_t \circ H_1\}) \le \bar{L}_p(X) + 2\pi \underline{L}_p(X) + \epsilon/2. \tag{e 15.61}$$

Moreover,

$$\overline{\text{Length}}(\{\pi_t \circ H_2\}) \le \bar{L}_p(X) + \epsilon/4. \tag{e 15.62}$$

It follows from (e 15.53), (e 15.57), (e 15.42) and the construction that

$$\|[\pi_t \circ H_1(f), u_t]\| < \epsilon \text{ for all } f \in \mathcal{F} \text{ and } t \in [0, 1], \tag{e 15.63}$$

and

$$\|[\pi_t \circ H_2, u_{1+t}]\| < \epsilon \text{ for all } f \in \mathcal{F} \text{ and } t \in [0, 1]. \tag{e 15.64}$$

It is then easy to check that $\{u_t : t \in [0, 2]\}$, H_1 and H_2 meet the requirements. □

The following is an improved version of the original Super Homotopy Lemma of [**3**] for purely infinite simple C^*-algebras. One notes that the length of $\{U_t\}$ and that of $\{V_t\}$ are considerably shortened.

Corollary 15.3. *For any $\epsilon > 0$, there exists $\delta > 0$ satisfying the following: Suppose that A is a unital purely infinite simple C^*-algebra and $u_1, u_2, v_1, v_2 \in U(A)$ are four unitaries. Suppose that*

$$\|[u_1, v_1]\| < \delta, \quad \|[u_2, v_2]\| < \delta,$$
$$[u_1] = [u_2], \quad [v_1] = [v_2] \text{ and } \text{bott}_1(u_1, v_1) = \text{bott}_1(u_2, v_2).$$

Then there exist two continuous paths of unitaries $\{U_t : t \in [0, 1]\}$ and $\{V_t : t \in [0, 1]\}$ of A such that

$$U_0 = u_1, \quad U_1 = u_2, \quad V_0 = v_1, \quad V_1 = v_2,$$
$$\|[U_t, V_t]\| < \epsilon \text{ for all } t \in [0, 1] \text{ and}$$
$$\text{Length}(\{U_t\}) \le 4\pi + \epsilon \text{ and } \text{Length}(\{V_t\}) \le 4\pi + \epsilon.$$

Moreover, if $sp(u_1) = sp(u_2) = S^1$, then the length of $\{U_t\}$ can be shortened to $2\pi + \epsilon$.

Proof. When $sp(u_1) = sp(u_2) = S^1$, the corollary follows immediately from Theorem 15.1. Here $X = S^1$. The subset $\mathcal{F}$ should contain $1_{C(X)}$ and z. Note that the function z has the Lipschitz constant $\bar{D}_z = 1$. Note also that if we take $\{u_t\}$ for $\{V_t\}$ and $U_t = \pi_t \circ H(z \otimes 1)$, then the pair will almost do the job. It is standard to bridge the ends of the path to u_1 and u_2 within an arbitrarily small error (both in norm and length).

For the general case, we apply Theorem 15.2. Note again that it is standard to bridge the the arbitrarily small gap between two paths of unitaries. □

16. Super Homotopy Lemma — the finite case

Lemma 16.1. *Let X be a finite CW complex and $\mathcal{F} \subset C(X)$ be a finite subset. Let $\epsilon > 0$ and $\gamma > 0$, there exits $\delta > 0$ and a finite subset $\mathcal{G} \subset C(X)$ satisfying the following:*

Suppose that A is a unital separable simple C^*-algebra with tracial rank zero and that $h : C(X) \to A$ is a unital homomorphism and $u \in A$ is a unitary such that

$$\|[h(g), u]\| < \delta \text{ for all } g \in \mathcal{G}. \tag{e 16.1}$$

Then, there is a projection $p \in A$, a unital homomorphism $h_1 : C(X) \to (1-p)A(1-p)$, a unital homomorphism $\Phi : C(X \otimes S^1) \to pAp$ with finite dimensional range and a unitary $v \in (1-p)A(1-p)$ such that

$$\|u - v \oplus \Phi(1 \otimes z)\| < \epsilon, \quad \|[v, h_1(f)]\| < \epsilon,$$
$$\|h(f) - (h_1(f) \oplus \Phi(f \otimes 1))\| < \epsilon \tag{e 16.2}$$

for all $f \in \mathcal{F}$ and

$$\tau(1-p) < \gamma \text{ for all } \tau \in T(A). \tag{e 16.3}$$

Proof. Let $\epsilon > 0$, $\gamma > 0$ and a finite subset $\mathcal{F} \subset C(X)$ be given. We may assume that $1_{C(X)} \in \mathcal{F}$.

Let $\delta > 0$ and a finite subset $\mathcal{G} \subset C(X)$ be as required by 5.4 corresponding to $\epsilon/4$ and $\gamma/2$ and $\mathcal{F}_1$.

We may assume that $\mathcal{F} \subset \mathcal{G}$ and $\delta < \epsilon/16$.

Let $\eta_1 = \sigma_{X,\mathcal{F},\epsilon/32}$. We may assume that $\eta_1 < 1$. Now suppose that h and u are as in the statement of the lemma for the above δ and $\mathcal{G}$. Suppose that $F \subset X$ is the spectrum of h.

Let $\{x_1, x_2, ..., x_n\} \subset F$ be an $\eta_1/4$-dense subset. Suppose that $O_i \cap O_j = \emptyset$ for $i \neq j$, where

$$O_i = \{x \in X : \text{dist}(x, x_i) < \eta_1/2s\}, \quad i = 1, 2, ..., n,$$

where $s \geq 1$ is an integer.

Let $g_i \in C(X)$ be such that $0 \leq g_i(x) \leq 1$, $g_i(x) = 1$ if $x \in O(x_i, \eta_1/4s)$ and $g_i(x) = 0$ if $x \notin O_i$, $i = 1, 2, ..., n$.

Put $\mathcal{F}_1 = \mathcal{F} \cup \{g_i : i = 1, 2, ..., n\}$ and put

$$Q_i = O(x_i, \eta_1/4s), \quad i = 1, 2, ..., n.$$

Let

$$\inf\{\mu_{\tau \circ h}(Q_i) : \tau \in T(A), i = 1, 2, ..., n\} \geq \sigma_1 \eta_1 \tag{e 16.4}$$

for some $\sigma_1 < 1/2s$.

Since F is compact, there are finitely many $y_1, y_2, ..., y_K \in F$ such that

$$\cup_{k=1}^K O(y_k, \eta_1/4) \supset F.$$

Denote $\Omega_k = \{x \in X : \text{dist}(x, y_k) \leq \eta_1/4\}$ and

$$Y = \cup_{k=1}^K \Omega_k.$$

Then Y is a finite CW complex and $F \subset Y \subset X$. Moreover, $\{x_1, x_2, ..., x_n\}$ is $\eta_1/2$ -dense in Y. Let $s : C(X) \to C(Y)$ be the quotient map and let $\bar{h} : C(Y) \to A$ be such that $\bar{h} \circ s = h$. Put

$$\bar{O}_i = \{y \in Y : \text{dist}(x_i, y) < \eta_1/2s\}, \quad i = 1, 2, ..., n.$$

Choose an integer L such that $1/L < \gamma/2$ and let $\sigma_2 = \frac{\sigma_1}{2(L+1)}$.

Let $\delta_1 > 0$ and $\mathcal{G}_1 \subset C(Y)$ be a finite subset and $\mathcal{P} \subset \underline{K}(C(Y))$ be a finite subset required by 13.6 corresponding to $\epsilon/2$, $s(\mathcal{F})$, σ_2 (and η_1) above. Let $\mathcal{G}_2 \subset C(X)$ be a finite subset such that $s(\mathcal{G}_2) \supset \mathcal{G}_1$.

Let $\delta_2 = \min\{\frac{\delta_1}{2}, \frac{\sigma_2 \cdot \eta_1}{2}\}$. We may assume that

$$[L_1]|_{\mathcal{P}} = [L_2]|_{\mathcal{P}}$$

for any pair of δ_2-$\mathcal{G}_1$-multiplicative contractive completely positive linear maps $L_1, L_2 : C(Y) \to A$ provided that

$$L_1 \approx_{\delta_2} L_2 \quad \text{on } \mathcal{G}_1.$$

By applying 5.4, there is a projection $q \in A$ and a unital homomorphism $\Psi : C(X \otimes S^1) \to qAq$ with finite dimensional range and a unitary $v \in (1-q)A(1-q)$ such that

$$\|u - v_1 \oplus \Psi(1 \otimes z)\| < \epsilon/2, \tag{e 16.5}$$

$$\|h(f)g(u) - ((1-q)h(f)g(u)(1-q) \oplus \Psi(f \otimes g)))\| < \epsilon/2 \tag{e 16.6}$$

for all $f \in \mathcal{F}_1$ and $g \in S$, where $S = \{1_{C(S^1)}, z\}$, and

$$\tau(1-q) < \gamma/2 \quad \text{for all } \tau \in T(A). \tag{e 16.7}$$

Moreover,

$$\|h(f) - ((1-q)h(f)(1-q) + \Psi(f \otimes 1))\| < \delta_2/2 \quad \text{and} \tag{e 16.8}$$

$$\|[(1-p), h(f)]\| < \delta_2/2 \quad \text{for all } f \in \mathcal{G}_2. \tag{e 16.9}$$

In particular, $(1-q)h(1-q)$ is $\delta_2/2$-$\mathcal{G}_2$-multiplicative.

We may write

$$\Psi(f \otimes g) = \sum_{k=1}^{m} (\sum_{j=1}^{n(k)} f(\xi_k) g(t_{k,j}) q_{k,j}) \tag{e 16.10}$$

for all $f \in C(X)$ and $g \in C(S^1)$, where $\xi_k \in X$, $t_{k,j} \in S^1$ and where $\{q_{k,j} : k, j\}$ is a set of mutually orthogonal projections with $\sum_{k=1}^{m} \sum_{j=1}^{n(k)} q_{k,j} = q$.

Note that the spectrum of h is F. Let $h' : C(F) \to A$ be the monomorphism induced by h. By applying 5.4 to h' instead of applying to h above, we may assume that $\xi_k \in F$, $k = 1, 2, ...m$.

For each j and k, there are mutually orthogonal projections $p_{k,j,l}$, $l = 1, 2, ..., L+1$ in A such that

$$q_{k,j} = \sum_{l=1}^{L+1} p_{k,j,l}, \quad [p_{k,j,l}] = [p_{k,j,1}], \tag{e 16.11}$$

$l = 1, 2, ..., L$ and $[p_{k,j,L+1}] \le [p_{k,j,1}]$. This can be easily arranged since A has tracial rank zero but it also follows from 9.4.

Put $p_{k,j} = \sum_{l=2}^{L+1} p_{k,j,l}$, $j = 1, 2, ..., n(k)$ and $k = 1, 2, ..., m$ and put

$$p = \sum_{k=1}^{m} \sum_{j=1}^{n(k)} p_{k,j}.$$

Define $\Phi : C(X \times S^1) \to pAp$ by

$$\Phi(f) = \sum_{k=1}^{m} \sum_{j=1}^{n(k)} f(\xi_k) g(t_{k,j}) p_{k,j} \text{ for all } f \in C(X) \text{ and } g \in C(S^1) \tag{e 16.12}$$

and define $\psi : C(X) \to (1-p)A(1-p)$ by

$$\psi(f) = (1-q)h(f)(1-q) + \sum_{k=1}^{m} f(\xi_k)(\sum_{j=1}^{n(k)} p_{k,j,1}) \tag{e 16.13}$$

for all $f \in C(X)$. Note that ψ is $\delta_2/2$-$\mathcal{G}_1$-multiplicative. Moreover, $\xi_k \in F \subset Y$, $k = 1, 2, ..., m$. Define $\psi_1 : C(Y) \to (1-p)A(1-p)$ by

$$\psi_1(f) = (1-q)\bar{h}(f)(1-p) + \sum_{k=1}^{m} f(\xi_k)(\sum_{j=1}^{n(k)} p_{k,j,1}) \tag{e 16.14}$$

for all $f \in C(Y)$.

Define $\phi : C(Y) \to pAp$ by

$$\phi(f) = \sum_{k=1}^{m} f(\xi_k)(\sum_{j=1}^{n(k)} p_{k,j})$$

for all $f \in C(Y)$. Write $Y = \sqcup_{k=1}^{m} Y_k$, as a disjoint union of finitely many path connected components, where each Y_k is a connected finite CW complex. Write $Z_k = Y_k \setminus \{y_k\}$, where $y_k \in Y_k$ is a point.

Note that

$$[\phi|_{C_0(\sqcup_k Z_k)}] = 0 \text{ in } KL(C_0(\sqcup_k Z_k), A). \tag{e 16.15}$$

Let $E_k \in C(Y)$ be the projection corresponding to Y_k, $k = 1, 2, ..., m$. It follows 4.2 that there is a unital homomorphism $h_{00} : C(Y) \to (1-p)A(1-p)$ such that

$$[h_{00}|_{C_0(\sqcup_k Z_k)}] = [h|_{C_0(\sqcup_k Z_k)}] \text{ in } KL(C_0(Y_X), A) \tag{e 16.16}$$

and

$$[h_{00}(E_k)] = [\psi_1(E_k)] \text{ in } K_0(A), k = 1, 2, ..., m. \tag{e 16.17}$$

It follows from (e 16.8) that

$$[h_{00}]|_{\mathcal{P}} = [\psi_1]|_{\mathcal{P}}. \tag{e 16.18}$$

On the other hand, by (e 16.8) again,

$$\|h(f) - (\psi(f) - \Phi(f \otimes 1))\| < \delta_2/2 \text{ for all } f \in \mathcal{G}_2. \tag{e 16.19}$$

We estimate, by (e 16.19) and (e 16.4) that

$$\begin{aligned}
\tau(\psi_1 \circ s(g_i)) &\geq \tau(h(g_i)) - \tau(\phi(s(g_i))) - \delta_2/2 && \text{(e 16.20)}\\
&\geq \tau(h(g_i)) - \frac{L\tau(\Phi(g_i))}{L+1} - \frac{\sigma_1 \cdot \eta_1}{4(L+1)} && \text{(e 16.21)}\\
&\geq \frac{\tau(h(g_i))}{L+1} - \delta_2/2 - \frac{\sigma_1 \cdot \eta_1}{4(L+1)} && \text{(e 16.22)}\\
&\geq \frac{\sigma_1 \cdot \eta_1}{L+1} - \frac{\sigma_1 \cdot \eta_1}{2(L+1)} && \text{(e 16.23)}\\
&= \frac{\sigma_1 \cdot \eta_1}{2(L+1)} \geq \sigma_2 \cdot \eta_1, && \text{(e 16.24)}
\end{aligned}$$

$i = 1, 2, ..., n$. It follows that

$$\mu_{\tau\circ\psi_1}(\bar{O}_i) \geq \sigma_2 \cdot \eta_1 \text{ for all } \tau \in T(A), \tag{e 16.25}$$

$i = 1, 2, ..., n$. Therefore

$$\mu_{t\circ\psi_1}(\bar{O}_i) \geq \sigma_2 \cdot \eta_1 \text{ for all } t \in T((1-p)A(1-p)).$$

By applying 13.6 to ψ_1, we obtain a unital monomorphism $\bar{h}_1 : C(Y) \to (1-p)A(1-qp)$ such that

$$\psi_1 \approx_{\epsilon/2} \bar{h}_1 \text{ on } s(\mathcal{F}). \tag{e 16.26}$$

Define $h_1 = \bar{h}_1 \circ s$. Then (e 16.26) becomes

$$\psi \approx_{\epsilon/2} h_1 \text{ on } \mathcal{F}. \tag{e 16.27}$$

It follows from (e 16.27) and (e 16.8) that

$$h \approx_\epsilon h_1 \oplus \Phi \text{ on } \mathcal{F}. \tag{e 16.28}$$

Put $v = v_1 \oplus \sum_{k=1}^n \sum_{j=1}^{n(k)} t_{k,j}p_{k,j,1}$. It is a unitary in $(1-p)A(1-p)$. Moreover,

$$\|u - v \oplus \Phi(1 \otimes z)\| < \epsilon.$$

The other inequality in (e 16.2) also follows. Finally,

$$\tau(1-p) = \tau(1-q) + \tau(\sum_{k=1}^{m}\sum_{j=1}^{n(k)} p_{k,j,1}) \tag{e 16.29}$$

$$< \gamma/2 + \frac{1}{L} < \gamma \tag{e 16.30}$$

for all $\tau \in T(A)$. □

Theorem 16.2. *Let X be a connected finite CW complex, let $\epsilon > 0$ and $\mathcal{F} \subset C(X)$ be a finite subset. Then, there exists $\delta > 0$ and a finite subset $\mathcal{G} \subset C(X)$ satisfying the following:*

Suppose that A is a unital separable simple C^*-algebra with tracial rank zero and suppose that $h_1, h_2 : C(X) \to A$ are two unital homomorphisms such that

$$[h_1] = [h_2] \text{ in } KL(C(X), A) \text{ and} \tag{e 16.31}$$

if there are unitaries $u, v \in A$ such that

$$\|[h_1(f), u]\| < \delta, \ \|[h_2(f), v]\| < \delta \text{ for all } f \in \mathcal{G} \text{ and} \tag{e 16.32}$$

$$\mathrm{Bott}(h_1, u) = \mathrm{Bott}(h_2, v), \tag{e 16.33}$$

then there exist a unital homomorphism $H_i : C(X) \to C([0,1], A)$ $(i = 1, 2)$ and a continuous rectifiable path $\{u_t : t \in [0,2]\}$ such that

$$u_0 = u, \ u_2 = v, \ \pi_0 \circ H_1 \approx_\epsilon h_1, \ \pi_1 \circ H_1 \approx_\epsilon \pi_0 \circ H_2 \text{ and } \pi_1 \circ H_2 \approx_\epsilon h_2 \tag{e 16.34}$$

on $\mathcal{F}$, and

$$\|[u_t, \pi_t \circ H_1(f)]\| < \epsilon \text{ and } \|[u_{t+1}, \pi_t \circ H_2(f)]\| < \epsilon \tag{e 16.35}$$

for all $t \in [0,1]$ and for all $f \in \mathcal{F}$. Moreover,

$$\overline{\text{Length}}(\{\pi_t \circ H_1 : t \in [0,2/3]\}) \le L(X) + \epsilon/2 \tag{e 16.36}$$

$$\text{Length}(\{\pi_t \circ H_1 : t \in [2/3,1]\}) \le 2\pi + \frac{\epsilon}{2(1+L(X))}, \tag{e 16.37}$$

$$\overline{\text{Length}}(\{\pi_t \circ H_1\}) \le L(X) + 2\pi \underline{L}_p(X) + \epsilon, \tag{e 16.38}$$

$$\overline{\text{Length}}(\{\pi_t \circ H_2\}) \le L(X) + \epsilon \text{ and} \tag{e 16.39}$$

$$\text{Length}(\{u_t\}) \le 4\pi + \epsilon. \tag{e 16.40}$$

PROOF. Let $\epsilon > 0$ and $\mathcal{F} \subset C(X)$ be a finite subset. Let

$$\mathcal{F}_1 = \{f \otimes g \in C(X \times S^1) : f \in \mathcal{F}, g = 1_{C(S^1)}, g = z\}.$$

We may assume that $\mathcal{F}$ is in the unit ball of $C(X)$.

Let $\eta_1 = \sigma_{X,\mathcal{F},\epsilon/32}$ and $\eta_2 = \min\{\sigma_{X\times S^1,\mathcal{F}_1,\epsilon/32}, \eta_1\}$. Let $\{x_1, x_2, ..., x_m\}$ be $\eta_1/4$-dense in X. Let $\{t_1, t_2, ..., t_l\}$ be l points on the unit circle which divides the unit circle into l even arcs. Moreover, we assume that $8m\pi/l < \min\{\epsilon/4, \eta_2/2\}$. Choose $s \ge 1$ such that $O(x_i \times t_j) \cap O(x_{i'} \times t_{j'}) = \emptyset$ if $i \ne i'$ or $j \ne j'$, where

$$O(x_i \times t_j) = \{x \times t \in X \times S^1 : \text{dist}(x, x_i) < \eta_1/2s \text{ and } \text{dist}(t, t_j) < \pi/sl\}.$$

Choose $0 < \sigma_1 < 1/2ml\eta_2$ and $\sigma_1 < 1/2s$.

Let $\mathcal{G}_1 \subset C(X \times S^1)$ be a finite subset, $\delta_1 > 0$, $1 > \gamma > 0$ and $\mathcal{P} \subset \underline{K}(C(X \times S^1))$ be as required by Theorem 4.6 of [**33**] corresponding to $\epsilon/4$, $\mathcal{F}_1$ and σ_1 above.

Let $\mathcal{P}_1 \subset \underline{K}(C(X))$ be a finite subset such that

$$\boldsymbol{\beta}(\mathcal{P}_1) \supset \hat{\boldsymbol{\beta}}(\mathcal{P}).$$

Let $\mathcal{P}_2 = \mathcal{P} \cup \{(\text{id} - \hat{\boldsymbol{\beta}})(\mathcal{P}), \hat{\boldsymbol{\beta}}(\mathcal{P})\}$.

We may also assume that

$$[L_1]|_{\mathcal{P}_2} = [L_2]|_{\mathcal{P}_2}$$

for any pair of δ_1-$\mathcal{G}_1$-multiplicative contractive completely positive linear maps $L_1, L_2 : C(X \times S^1) \to A$, provided that

$$L_1 \approx_{\delta_1} L_2 \text{ on } \mathcal{G}_1.$$

Without loss of generality, we may assume that $\delta_1 < \sin(\epsilon/4)$ and we may assume that $\mathcal{G}_1 = \mathcal{G}_1' \otimes S$, where $\mathcal{G}_1' \subset C(X)$, where $S = \{1_{C(S^1)}, z\}$. Moreover, we may assume that both sets are in the unit balls of $C(X)$ and $C(S^1)$, respectively.

Without loss of generality, we may also assume that $\gamma < 1/2$. Let $\delta_2 > 0$ (in place of δ) and $\mathcal{G}_2 \subset C(X)$ (in place of $\mathcal{F}_1$) be a finite subset required by 2.8 for $\delta_1/2$ (in place of ϵ) and $\mathcal{G}_1'$ (in place of $\mathcal{F}_0$). We may assume that $\delta_2 < \delta_1/2$ and $\mathcal{G}_1' \subset \mathcal{G}_2$.

Let $\delta > 0$ and $\mathcal{G}_3 \subset C(X)$ be a finite subset required in 16.1 for $\delta_2/2$, $\gamma/4$ and $\mathcal{G}_2$. We may assume that $\mathcal{G}_2 \subset \mathcal{G}_3$ and $\delta_3 < \delta_2/2$.

Let $\mathcal{G} = \mathcal{G}_3 \cup \mathcal{F}$.

Now suppose that

$$\|[u, h_1(g)]\| < \delta \text{ and } \|[v, h_2(g)]\| < \delta \tag{e 16.41}$$

for all $g \in \mathcal{G}$ and

$$[h_1] = [h_2] \text{ in } KL(C(X), A) \text{ and } \text{Bott}(h_1, u) = \text{Bott}(h_2, v). \tag{e 16.42}$$

In particular,

$$\mathrm{Bott}(h_1, u)|_{\mathcal{P}_1} = \mathrm{Bott}(h_2, v)|_{\mathcal{P}_1}. \tag{e 16.43}$$

By applying 16.1, we obtain projections $p_i \in A$ and unital homomorphisms $h_{1,i} : C(X) \to (1-p_i)A(1-p_i)$, unital homomorphisms $\Phi_i : C(X \times S^1) \to p_iAp_i$ and unitaries $w_i' \in (1-p_i)A(1-p_i)$ such that

$$\|u - w_1 \oplus \Phi_1(1 \otimes z)\| < \delta_2/2, \|v - w_2 \oplus \Phi_2(1 \otimes z)\| < \delta_2/2, \tag{e 16.44}$$

$$\|[w_i, h_i(f)]\| < \delta_2/2 \text{ and } h_i \approx_{\delta_2/2} h_{1,i} \oplus \phi_i \text{ on } \mathcal{G}_2, \tag{e 16.45}$$

where $\phi_i(f) = \Phi_i(f \otimes 1)$ for $f \in C(X)$, $i = 1, 2$. Moreover,

$$\tau(1-p_i) < \gamma/4 \text{ for all } \tau \in T(A), \quad i = 1, 2. \tag{e 16.46}$$

Using Zhang's Riesz interpolation as in the proof of 5.9, we may assume (by replacing $\gamma/4$ by $\gamma/2$, if necessary, in (e 16.45)) that

$$[p_1] = [p_2].$$

Define $\Psi^{(0)} : C(X \times S^1) \to p_1Ap_1$ by

$$\Psi^{(0)}(f) = \sum_{k=1}^{m}\sum_{j=1}^{l} f(x_k)g(t_j)e_{k,j} \text{ for all } f \in C(X) \text{ and } g \in C(S^1),$$

where $\{e_{k,j} : k, j\}$ is a set of mutually orthogonal projections in p_1Ap_1 such that

$$\tau(e_{k,j}) \geq \frac{\tau(1-p_1)}{2ml} \geq \sigma_1 \cdot \eta_2 \text{ for all } \tau \in T(A). \tag{e 16.47}$$

$(i = 1, 2)$.

It follows from 5.8 that there is unital commutative finite dimensional C^*-subalgebra $B_i \subset p_iAp_i$ which contains the image of $\Phi_i(C(X \times S^1))$ and a set of mutually orthogonal projections $e_{k,j}^{(i)}$ such that

$$[e_{k,j}^{(i)}] = [e_{k,j}], \quad k = 1, 2, ..., m, j = 1, 2, ..., l, \tag{e 16.48}$$

and $i = 1, 2$.

Define $\Psi_{0,i} : C(X \times S^1) \to B_i$ by

$$\Psi_{0,i}(f \otimes g) = \sum_{k=1}^{m}\sum_{j=1}^{l} f(x_k)g(t_j)e_{k,j}^{(i)} \tag{e 16.49}$$

for all $f \in C(X)$ and $g \in C(S^1)$.

By 12.11, there is a unital homomorphism $H_i' : C(X) \to C([0,1], B_i)$ and a continuous path of unitaries $w_{i,t} : t \in [0,1]$ in p_iAp_i such that

$$w_{i,0} = \Phi_i(1 \otimes z), \quad w_{i,1} = \Psi_{0,i}(1 \otimes z), \tag{e 16.50}$$

$$\pi_0 \circ H_i'(f) = \Phi_i(f \otimes 1), \pi_1 \circ H_i'(f) = \Psi_{0,i}(f \otimes 1) \text{ and} \tag{e 16.51}$$

$$[\pi_t \circ H_i'(f), w_{i,t}] = 0 \text{ for all } t \in [0,1] \text{ and } \text{ for all } f \in C(X), \tag{e 16.52}$$

$i = 1, 2$. Moreover,

$$\overline{\mathrm{Length}}(\{\pi_t \circ H_i'\}) \leq L(X) + \epsilon/2 \text{ and } \mathrm{Length}(\{w_{i,t}\}) \leq \pi, \tag{e 16.53}$$

$i = 1, 2$.

From (e 16.44) and (e 16.45), by applying 2.8, we obtain a $\delta_1/2$-$\mathcal{G}_1'$ -multiplicative contractive completely positive linear map $\psi_i' : C(X \times S^1) \to A$ such that

$$\|\psi_i'(f \otimes g) - h_{1,i}(f)g(w_i)\| < \delta_1/2 \tag{e 16.54}$$

for all $f \in \mathcal{G}_1'$ and $g \in S$, $i = 1, 2$. Define $\psi_i : C(X \times S^1) \to A$ by

$$\psi_i(f \otimes g) = \psi_i'(f \otimes g) \oplus \Phi_{0,i}(f \otimes g) \tag{e 16.55}$$

for all $f \in C(X)$ and $g \in C(S^1)$. Moreover, by the fact that $[p_1] = [p_2]$, (e 16.42), (e 16.43), (e 16.44), (e 16.45) and (e 16.48), we compute that

$$[\psi_1]|_{\mathcal{P}} = [\psi_2]|_{\mathcal{P}}. \tag{e 16.56}$$

By (e 16.47), (e 16.49) and (e 16.48),

$$\mu_{\tau \circ \phi_i}(O(x_k \times t_j)) \geq \sigma_2 \cdot \eta_2 \text{ for all } \tau \in T(A), \tag{e 16.57}$$

$k = 1, 2, .., m$, $j = 1, 2, ..., l$ and $i = 1, 2$. Furthermore, since $[e_{k,j}^{(1)}] = [e_{k,j}^{(2)}]$, $k = 1, 2, ..., m$, $j = 1, 2, ..., l$, we compute that

$$|\tau \circ \psi_1(f) - \tau \circ \psi_2(f)| < \gamma \text{ for all } f \in \mathcal{G}_1. \tag{e 16.58}$$

Now, from the choice of δ_1, γ, $\mathcal{G}_1$, by applying Theorem 4.6 of [**33**], we obtain a unitary $U \in A$ such that

$$\mathrm{ad}\, U \circ \psi_1 \approx_{\epsilon/4} \psi_2 \text{ on } \mathcal{F}_1. \tag{e 16.59}$$

There is a continuous path of unitaries $\{U_t : t \in [0, 1]\}$ such that

$$U_0 = 1, \;\; U_1 = U \text{ and } \mathrm{Length}(\{U_t\}) \leq \pi + \frac{\epsilon}{4(1 + L(X))}. \tag{e 16.60}$$

Define $H_1 : C(X) \to C([0, 1], A)$ by, for $f \in C(X)$,

$$\pi_t \circ H_1(f) = \begin{cases} h_{1,1}(f) \oplus \phi_1(f) & \text{if } t \in [0, 1/3]; \\ h_{1,1}(f) \oplus \pi_{3(t-1/3)} \circ H_1'(f) & \text{if } t \in (1/3, 2/3]; \\ U_{3(t-2/3)}^*(h_{1,1}(f) \oplus \Psi_{0,1}(f \otimes 1))U_{3(t-2/3)} & \text{if } t \in (2/3, 1]. \end{cases} \tag{e 16.61}$$

Define $H_2 : C(X) \to C([0, 1], A)$ by

$$\pi_t \circ H_2 = \begin{cases} h_{1,2} \oplus \pi_{2(1/2-t)} \circ H_2' & \text{if } t \in [0, 1/2]; \\ h_{1,2} \oplus \phi_2 & \text{if } t \in (1/2, 1]. \end{cases} \tag{e 16.62}$$

Then, by (e 16.45) and (e 16.59),

$$\pi_0 \circ H_1 \approx_\epsilon h_1, \;\; \pi_1 \circ H_1 \approx_\epsilon \pi_0 \circ H_2 \text{ and } \pi_1 \circ H_2 \approx_\epsilon h_2 \tag{e 16.63}$$

on $\mathcal{F}$. Moreover, by (e 16.53) and (e 16.60)

$$\overline{\mathrm{Length}}(\{\pi_t \circ H_1 : t \in [0, 2/3]\}) \leq L(X) + \epsilon/2, \tag{e 16.64}$$

$$\mathrm{Length}(\{\pi_t \circ H_1 : t \in [2/3, 1]\} \leq 2\pi + \frac{\epsilon}{2(1 + L(X))} \tag{e 16.65}$$

and

$$\overline{\mathrm{Length}}(\{\pi_t \circ H_2\}) \leq L(X) + \epsilon. \tag{e 16.66}$$

So, by 12.8,

$$\overline{\mathrm{Length}}(\{\pi_t \circ H_1\}) \leq L(X) + 2\pi \underline{L}_p(X) + \epsilon. \tag{e 16.67}$$

By (e 16.44), since $\delta_1 < \sin(\epsilon/4)$, there is $a_i \in A_{s.a}$ with $\|a_i\| < \epsilon/4$ such that

$$u^*(w_1 \oplus \Phi_1(1 \otimes z)) = \exp(ia_1) \text{ and } v^*(w_2 \oplus \Phi_2(1 \otimes z)) = \exp(ia_2).$$

Define

$$(e\,16.68) \qquad u_t = \begin{cases} u\exp(i3ta_1) & \text{if } t \in [0, \frac{1}{3}]; \\ w_1 \oplus w_{1,3(t-1/3)} & \text{if } t \in (\frac{1}{3}, \frac{2}{3}]; \\ U^*_{3(t-1/3)} \circ (w_1 \oplus w_{1,1})U_{3(t-1/3)} & \text{if } t \in (\frac{2}{3}, 1]; \\ w_1 \oplus w_{2,2(3/2-t)} & \text{if } t \in (1, \frac{3}{2}]; \\ v\exp(i2(2-t)a_2) & \text{if } t \in (\frac{3}{2}, 2]. \end{cases}$$

Then

$$u_0 = u, \quad u_2 = v \text{ and}$$
$$\text{Length}(\{u_t\}) \le \epsilon/4 + \pi + 2\pi + \pi + \epsilon/4 = 4\pi + \epsilon/2.$$

We also have

$$\|[u_t, \pi_t \circ H_1(f)]\| < \epsilon \text{ for all } f \in \mathcal{F}$$

and

$$\|[u_{t+1}, \pi_t \circ H_2(f)]\| < \epsilon \text{ for all } f \in \mathcal{F}$$

for $t \in [0, 1]$. □

We obtain the following improvement of the original Super Homotopy Lemma in the case that A is a unital separable simple C^*-algebra of tracial rank zero.

Corollary 16.3. *For any $\epsilon > 0$, there exists $\delta > 0$ satisfying the following: Suppose that A is a unital separable simple C^*-algebra with tracial rank zero and $u_1, u_2, v_1, v_2 \in U(A)$ are four unitaries. Suppose that*

$$\|[u_1, v_1]\| < \delta, \quad \|[u_2, v_2]\| < \delta,$$
$$[u_1] = [u_2], \quad [v_1] = [v_2] \text{ and } \text{bott}_1(u_1, v_1) = \text{bott}_1(u_2, v_2).$$

Then there exist two continuous paths of unitaries $\{U_t : t \in [0, 1]\}$ and $\{V_t : t \in [0, 1]\}$ of A such that

$$U_0 = u_1, \quad U_1 = u_2, \quad V_0 = v_1, \quad V_1 = v_2,$$
$$\|[U_t, V_t]\| < \epsilon \text{ for all } t \in [0, 1] \text{ and}$$
$$\text{Length}(\{U_t\}) \le 4\pi + \epsilon \text{ and } \text{Length}(\{V_t\}) \le 4\pi + \epsilon.$$

Remark 16.4. As stated in [**3**], the Basic Homotopy Lemma is not a special case of the Super Homotopy Lemma. It mentioned not only the estimate on the length of the paths but more importantly, the presence of two paths of unitaries in the Super Homotopy Lemma. But in the Basic Homotopy Lemma it is essential to keep one unitary fixed. In our cases also, in the Basic Homotopy Lemma, the map h remains fixed in the homotopy while 15.2 and 16.2 allow a path from h_1 to h_2. We would point out that beside the obvious difference in appearance of paths and the estimates of the length of paths, one additional essential difference appears: In the finite case, the constant δ in the Super Homotopy Lemma 16.2 is universal while in 8.3 δ depends on a measure distribution which as shown in 9.5 can not be universal. This phenomenon does not occur in the case that X is one-dimensional. This again shows the additional difficulties involved in the Basic Homotopy Lemma for higher dimensional spaces X. Moreover, in the Basic Homotopy Lemma, h is assumed to be a monomorphism and it fails when h is not assumed to be a monomorphism whenever $\dim X \geq 2$. However, in both 15.2 and 16.2, h_1 and h_2 are only assumed to be homomorphisms.

On the other hand, there are additional complications to connect two homomorphisms h_1 and h_2 as well as two unitaries while the Basic Homotopy Lemma considers only one unitary and one monomorphism. It should be noted that the Super Homotopy Lemma does not follow from 14.2 or 14.3, since $\pi_t \circ H_i$ is not a homomorphism.

It is also important to note that even in the case that $L(X) = \infty$, Theorems 13.1, 13.3, 13.5, 13.8 13.9, 14.1, 14.2, 14.3, 15.1, 15.2 and 16.2 remain valid. However, the lengths of these homotopy may be infinite.

CHAPTER 6

Postlude

17. Non-commutative cases

Definition 17.1. (i) Let X be a compact metric space, let r be a positive integer and let $A = M_r(C(X)) = C(X, M_r)$. Denote by tr the standard normalized trace on M_r. Let $\tau \in T(A)$. Then there is a probability Borel measure ν such that

$$\tau(f) = \int_X tr(f)d\nu \text{ for all } f \in C(X, M_r).$$

Define $\mu_\tau = r\nu$.

Suppose that B is a unital C^*-algebra and $h : A \to B$ is a unital homomorphism. Let $\{e_{i,j}\}$ be a system of matrix units for M_r. Identify $e_{1,1}$ with the (constant) rank one projection in A. Then $e_{1,1}Ae_{1,1} \cong C(X)$. Suppose that $t \in T(B)$. Let $\tau = t \circ h$. Define $h_1 = h|_{e_{1,1}Ae_{1,1}}$. Define $\tau_1(f) = r\tau \circ h_1(f)$ for $f \in C(X)$. Then τ_1 is a tracial state of $C(X)$. The associated measure is μ_τ.

(ii) Suppose that X is a connected finite CW complex with covering dimension d. Let $A = PM_m(C(X))P$, where P is a rank k ($\leq m$) projection in $M_m(C(X))$. Let $\tau \in T(A)$. Denote by tr the normalized trace on M_k. Then there exists a Borel probability measure ν on X such that

$$\tau(f) = \int_X tr(f)d\nu \text{ for all } f \in A.$$

Define $\mu_\tau = k\nu$.

Let B be a unital C^*-algebra, and let $h : A \to B$ be a unital homomorphism. There is a projection $e \in M_{d+1}(A)$ such that $e + 1_A$ is a trivial projection on X (see for example 8.12 of [**16**]). Therefore $(e + 1_A)M_{d+1}(A)(e + 1_A) \cong M_r(C(X))$ for some integer $r \geq 1$. Let $t = (\frac{r}{k(d+1)})(\tau \otimes Tr)$, where Tr is the standard trace on M_{d+1}. Then t gives a tracial state on $M_r(C(X))$. Using notation of (i), the measure μ_t on X is same as μ_τ defined above.

(iii) Let X be a (not necessary connected) finite CW complex with finite covering dimension and let $A = PM_m(C(X))P$, where P is a projection in $M_m(C(X))$. Then one may write $X = \sqcup_{i=1}^k X_i$, where each X_i is connected. So we may also write $A = \bigoplus_{i=1}^k P_iM_{m(k)}(C(X_i))P_i$, where $P_i \in M_{m(i)}(C(X_i))$ is a rank $r(i)$ projection. Put $A_i = P_iM_{m(k)}(C(X_i))P_i$, $i = 1, 2, ..., k$. Suppose that $\tau \in T(A)$. Put $\alpha_i = \tau(P_i)$ and $\tau_i = (1/\alpha_i)\tau|_{A_i}$. By applying (ii), one then obtains a probability Borel measure μ_{τ_i} on X_i. For any Borel subset $S \subset X$, write $S = \sqcup_{i=1}^k S_i$, where each $S_i \in X_i$ is a Borel subset. Define μ_τ by

$$\mu_\tau(S) = \sum_{i=1}^k \alpha_i \mu_{\tau_i}(S_i)$$

for all Borel subsets $S \subset X$.

Lemma 17.2. *Let X be a compact metric space and let $A = M_r(C(X))$, where $r \geq 1$ is an integer. Then, for any $\epsilon > 0$, any finite subset $\mathcal{F} \subset A$ and any increasing map $\Delta : (0,1) \to (0,1)$, there exists $\delta > 0$, a finite subset $\mathcal{G} \subset A$ and a finite subset $\mathcal{P} \subset \underline{K}(A)$ satisfying the following:*

Suppose that B is a unital separable simple C^*-algebra with tracial rank zero, $h : A \to B$ is a unital monomorphism and $u \in B$ is a unitary such that $\mu_{\tau \circ h}$ is Δ-distributed for all $\tau \in T(B)$,

$$\|[h(a), u]\| < \delta \text{ for all } a \in \mathcal{G} \text{ and } \mathrm{Bott}(h, u)|_{\mathcal{P}} = 0. \tag{e 17.1}$$

Then, there exists a continuous rectifiable path of unitaries $\{u_t : t \in [0,1]\}$ of B such that

$$u_0 = u, \ u_1 = 1_B \text{ and } \|[h(a), u_t]\| < \epsilon \text{ for all } a \in \mathcal{F} \text{ and } t \in [0,1]. \tag{e 17.2}$$

Moreover,

$$\mathrm{Length}(\{u_t\}) \leq 2\pi + \epsilon. \tag{e 17.3}$$

Proof. Let $\epsilon > 0$ and $\mathcal{F} \subset M_r(C(X))$ be given. There is a finite subset $\mathcal{F}_1 \subset C(X)$ such that

$$\mathcal{F} \subset \{(g_{i,j}) : g_{i,j} \in \mathcal{F}_1\}.$$

Let $\delta_1 > 0$, $\mathcal{G}_1 \subset C(X)$ be a finite subset and $\mathcal{P}_1 \subset \underline{K}(C(X)) = \underline{K}(A)$ be a finite subset as required by 8.1 associated with $\epsilon/2$, $\mathcal{F}_1$ and Δ. We may assume that

$$[L_1]|_{\mathcal{P}} = [L_2]|_{\mathcal{P}}$$

for any pair of δ_1-$\mathcal{G}_1$-multiplicative contractive completely positive linear maps $L_1, L_2 : C(X) \to A$, provided that

$$L_1 \approx_{\delta_1} L_2 \text{ on } \mathcal{G}_1.$$

We may assume that $\delta_1 < \epsilon/2$.

Let $\delta_2 > 0$ and $\mathcal{G}_2 \subset A$ be a finite subset required by 2.8 for $\delta_1/2$ and $\mathcal{G}_1$. We may assume that $\delta_2 < \min\{\delta_1/2, \epsilon/2\}$ and $\mathcal{G}_2 \supset \mathcal{G}_1 \cup \mathcal{F}_1$.

Let $\{e_{i,j} : i, j = 1, 2, ..., r\}$ be a system of matrix units for M_r.

It is easy to see that there is $\delta > 0$ and a finite subset $\mathcal{G} \subset A$ satisfying the following: if B is a unital C^*-algebra, if $h' : A \to B$ is a unital homomorphism and if $u \in B$ is a unitary with

$$\|[h'(f), u]\| < \delta \text{ for all } f \in \mathcal{G},$$

then, there exists a unitary $v \in B$ such that

$$\|u - v\| < \delta_2/2, \ h'(e_{1,1})v = vh'(e_{1,1}) \text{ and } \|[h'(g), v]\| < \delta_2/2$$

for all $f \in \mathcal{G}_3$, where

$$\mathcal{G}_3 = \{(g_{ij}) \in M_k(C(X)) : g_{i,j} \in \mathcal{G}_2\}.$$

Suppose that $h : A \to B$ is a monomorphism and $u \in A$ is a unitary satisfying the conditions in the lemma with δ and $\mathcal{G}$ as above. Define $h_1 : C(X) \to h(e_{1,1})Bh(e_{1,1})$ by $h_1 = h|_{e_{1,1}Ae_{1,1}}$. Then, 8.3 applies to h_1 and $e_{1,1}ve_{1,1}$.

One sees that we reduce the general case to the case that $r = 1$. $\square$

Lemma 17.3. *Let X be a compact metric space and let $A = M_r(C(X))$, where $r \geq 1$ is an integer. Then, for any $\epsilon > 0$, any finite subset $\mathcal{F} \subset A$, there exists $\delta > 0$, a finite subset $\mathcal{G} \subset A$ and a finite subset $\mathcal{P} \subset \underline{K}(A)$ satisfying the following:*

Suppose that B is a unital purely infinite simple C^*-algebra, $h : A \to B$ is a unital monomorphism and $u \in B$ is a unitary such that

$$\|[h(a), u]\| < \delta \text{ for all } a \in \mathcal{G} \text{ and } \mathrm{Bott}(h, u)|_{\mathcal{P}} = 0. \tag{e 17.4}$$

Then, there exists a continuous rectifiable path of unitaries $\{u_t : t \in [0,1]\}$ of B such that

$$u_0 = u, \;\; u_1 = 1_B \text{ and } \|[h(a), u_t]\| < \epsilon \text{ for all } a \in \mathcal{F} \text{ and } t \in [0,1]. \tag{e 17.5}$$

Moreover,

$$\mathrm{Length}(\{u_t\}) \leq 2\pi + \epsilon \tag{e 17.6}$$

PROOF. The proof of this lemma is almost identical to that of 17.2. We will apply 11.3. □

The essential significance in the following lemma is the estimate of the length of $\{v_t\}$. It should be noted that the proof would be much shorter if we allow the length to be bounded by $2L + \epsilon$.

Lemma 17.4. (1) *There is a positive number δ (with $1/2 > \delta > 0$) and there is a function $\eta : (0, 1/2) \to \mathbb{R}_+$ with $\lim_{t \to 0} \eta(t) = 0$ satisfying the following:*

Let A be a unital C^*-algebra and $\{u_t : t \in [0,1]\}$ be a path of unitaries in A with $u_1 = 1$ such that

$$\mathrm{Length}(\{u_t\}) \leq L.$$

Suppose that there is a non-zero projection $p \in A$ such that

$$\|[u_t, p]\| < \delta \text{ for all } t \in [0,1].$$

Then there exists a path of unitaries $\{v_t : t \in [0,1]\} \subset pAp$ and there exist partitions

$$0 = t_0 < t_1 \cdots < t_m = 1 \text{ and } 0 = s_0 < s_1 \cdots < s_m = 1$$

such that

$$\|pu_{t_i}p - v_{s_i}\| < \frac{\delta}{\sqrt{1-\delta}}, \;\; i = 1, 2, ..., m, \;\; v_1 = p \text{ and}$$
$$\mathrm{Length}(\{v_t\}) \leq L + (7L\eta(\delta)),$$

where $m = [\frac{6L}{\pi}] + 1$.

(2) *For any $\epsilon > 0$ and any $M > 0$. There exists $\delta > 0$ satisfying the following:*

Let A be a unital C^*-algebra and $\{u_t : t \in [0,1]\}$ be a path of unitaries in A with $u_1 = 1$ such that

$$\mathrm{Length}(\{u_t\}) \leq L.$$

Suppose that there is a non-zero projection $p \in A$ and a self-adjoint subset $\mathcal{G} \subset pAp$ with $p \in \mathcal{G}$ and $\|a\| \leq M$ for all $a \in \mathcal{G}$ such that

$$\|[u_t, a]\| < \delta \text{ for all } t \in [0,1].$$

Then there exists a path of unitaries $\{v_t : t \in [0,1]\} \subset pAp$ such that

$$\|pu_0p - v_0\| < \epsilon, \;\; v_1 = p,$$
$$\|[v_t, a]\| < \epsilon \text{ for all } a \in \mathcal{G} \text{ and}$$
$$\mathrm{Lengh}(\{v_t\}) \leq L + \epsilon.$$

PROOF. The proof is somewhat similar to that of 9.3. Fix $M > 0$. Put

$$S = \{e^{i\pi t} : t \in [-1/2, 1/2]\}.$$

Denote by $F : S \to [-1/2, 1/2]$ by $F(e^{i\pi t}) = t$. F is a continuous function. Let

$$S_1 = \{e^{i\pi t} : t \in [-1/6, 1/6]\}.$$

If U_1 and U_2 are two unitaries in a unital C^*-algebra A and $p \in A$ is a projection such that

$$\|[p, U_i]\| < \delta, \ \ i = 1, 2, \tag{e 17.7}$$

then

$$\|pU_ip|pU_ip|^{-1} - pU_ip\| < \frac{\delta}{\sqrt{1-\delta}}, \ \ i = 1, 2. \tag{e 17.8}$$

Here the inverse $|pU_ip|^{-1}$ is taken in pAp. Note that $pU_ip|pU_ip|^{-1}$ is a unitary in pAp. If $\mathrm{sp}(U_1U_2) \subset S_1$, then

$$\|1 - U_1U\| = \max\{|1 - e^{i\pi\theta}| : \theta \in [-1/6, 1/6]\} \leq 1/2. \tag{e 17.9}$$

We estimate that

(e 17.10)

$$\begin{aligned} &\|p - pU_1p|pU_1p|^{-1}pU_2p|pU_2p|^{-1}\| \\ &\leq \|p - pU_1U_2p\| + \|pU_1U_2p - pU_1p|pU_1p|^{-1}pU_2p|pU_2p|^{-1}\| \\ &< 1/2 + \|pU_1U_2p - pU_1pU_2p\| + \|pU_1pU_2p - pU_1p|pU_1p|^{-1}pU_2p|pU_2p|^{-1}\| \end{aligned}$$

(e 17.11)

$$< 1/2 + \delta + 2\frac{\delta}{1-\delta}.$$

There is $\delta_0 > 0$ such that if $\delta < \delta_0$,

$$1/2 + \delta + 2\frac{\delta}{1-\delta} < 1.$$

Thus when $\delta < \delta_0$, $\mathrm{sp}(U_1U_2) \subset S_1$ and when (e 17.7) holds,

$$\mathrm{sp}(pU_1p|pU_1p|^{-1}pU_2p|pU_2p|^{-1}) \subset S. \tag{e 17.12}$$

Fix $M > 0$.

Moreover, by choosing smaller $\delta_0 > 0$, when $0 < \delta < \delta_0$ and when (e 17.7) holds, there is a positive number $\eta(\delta)$ (for each δ) with

$$\lim_{t\to 0} \eta(t) = 0 \tag{e 17.13}$$

such that

$$\|\exp(i\pi F(pU_1p|pU_1p|^{-1}pU_2p|pU_2p|^{-1})) - p\exp(i\pi F(U_1U_2))p\| < \eta(\delta) \tag{e 17.14}$$

and at the same time

$$\|F(pU_1p|pU_1p|^{-1}pU_2p|pU_2p|^{-1}) - pF(U_1U_2)p\| < \eta(\delta) \tag{e 17.15}$$

and

$$\|[a, F(pU_1p|pU_1p|^{-1}pU_2p|pU_2p|^{-1})]\| < \eta(\delta), \tag{e 17.16}$$

if $\|[a, U_i]\| < \delta$, $a \in pAp$ and $\|a\| \leq M$ (cf. 2.5.11 of **[29]** and 2.6.10 of **[29]**). It should be noted that

$$\|F(pU_1pU_2p|pU_1pU_2p|^{-1})\| \leq \|F(U_1U_2)\| + \eta(\delta). \tag{e 17.17}$$

We will use these facts in the rest of the proof and assume that $0 < \delta < \delta_0$.

Let $0 = t_0 < t_1 < \cdots < t_m = b$ be a partition so that

$$\pi/7 \le \text{Length}(\{u_t : t \in [t_{i-1}, t_i]\}) \le \pi/6, \quad i = 1, 2, ..., m, \tag{e 17.18}$$

where $m = [\frac{6L}{\pi}] + 1$.

Put $u_i = u_{t_i}$ and $z_i = pu_ip|pu_ip|^{-1}$, where the inverse is taken in pAp. Then $\mathrm{sp}(u_{i-1}^*u_i) \subset S_1$. Define $\tilde{w}_i = z_{i-1}^* z_i$, $i = 1, 2, ..., m$.

If

$$\|[u_t, p]\| < \delta, \tag{e 17.19}$$

then (by (e 17.8)

$$\|pu_ip|pu_ip|^{-1} - pu_ip\| < \frac{\delta}{\sqrt{1-\delta}}, \tag{e 17.20}$$

$i = 1, 2, ..., m$. It follows from (e 17.12)that

$$sp(\tilde{w}_i) \subset S.$$

Moreover by (e 17.19), (e 17.12) and (e 17.14),

$$\begin{aligned} &\|z_{i-1}\exp(\sqrt{-1}\pi F(\tilde{w}_i)) - pu_ip\| \\ &\quad = \|z_{i-1}\exp(\sqrt{-1}\pi F(\tilde{w}_i)) - pu_{i-1}u_{i-1}^*u_ip\| \\ &\quad < \|z_{i-1}\exp(\sqrt{-1}\pi F(\tilde{w}_i)) - pu_{i-1}pu_{i-1}^*u_ip\| + \delta \\ &\quad < \|z_{i-1}\exp(\sqrt{-1}\pi F(\tilde{w}_i)) - pu_{i-1}p\exp(\sqrt{-1}\pi F(\tilde{w}_i))\| + \eta(\delta) + \delta \\ &\quad < \frac{\delta}{\sqrt{1-\delta}} + \eta(\delta) + \delta, \quad i = 1, 2, ..., m. \end{aligned} \tag{e 17.21}$$

Suppose that $\mathcal{G} \subset pAp$ is self-adjoint such that $\|a\| \le M$ for all $a \in \mathcal{G}$. Then, if $\delta < \delta_0$ and if

$$\|[a, u_t]\| < \delta \text{ for all } a \in \mathcal{G} \text{ and } t \in [0, 1],$$

then, by (e 17.16),

$$\|[a, F(\tilde{w}_i)]\| \le \eta(\delta) \text{ for all } a \in \mathcal{G}. \tag{e 17.22}$$

Thus, by (e 17.21), (e 17.13) and (e 17.20), a positive number δ depending on ϵ, M and F only, such that, if

$$\|[a, u_t]\| < \delta \text{ for all } a \in \mathcal{G} \text{ and } t \in [0, 1],$$

then,

$$\|[a, z_{i-1}\exp(\sqrt{-1}\,\pi t\, F(\tilde{w}_i))]\| < \epsilon \tag{e 17.23}$$

for all $t \in [0, 1]$ and for all $a \in \mathcal{G}$. Let

$$l_i = \text{Length}(\{u_t : t \in [t_{i-1}, t_i]\}), i = 1, 2, ..., m.$$

Now, define

$$v_0 = u_0, \ \ v_t = z_{i-1}(\exp(\sqrt{-1}\pi\frac{t - s_{i-1}}{s_i - s_{i-1}}F(\tilde{w}_i))) \text{ for all } s \in [s_{i-1}, s_i],$$

where $s_0 = 0$, $s_i = \sum_{j=1}^i l_j/L$, $i = 1, 2, ..., m$. It follows that

$$\|[a, v_t]\| < \epsilon \text{ for all } t \in [0, 1] \tag{e 17.24}$$

and for all $a \in \mathcal{G}$.

Clearly, if $t, t' \in [s_{i-1}, s_i]$, then (also using (e 17.17) and (e 17.18))

$$\begin{aligned} \|v_t - v_{t'}\| &= \|\exp(\sqrt{-1}\pi\frac{t-s_{i-1}}{s_i - s_{i-1}}F(\tilde{w}_i)) - \exp(\sqrt{-1}\pi\frac{t'-s_{i-1}}{s_i - s_{i-1}}F(\tilde{w}_i))\| \\ &\le \|\pi F(\tilde{w}_i)\|\frac{|t-t'|}{l_i/L} \le L\frac{(l_i + \eta(\delta)\cdot\pi)}{l_i}|t-t'| \\ &= (L + 7L\eta(\delta))|t-t'|. \qquad \text{(e 17.25)} \end{aligned}$$

One then computes that

$$\|v_t - v_{t'}\| \le (L + 7L\eta(\delta))|t-t'| \text{ for all } t, t' \in [0,1]. \qquad \text{(e 17.26)}$$

The lemma follows from the fact that $\lim_{\delta\to 0}\eta(\delta) = 0$. □

Lemma 17.5. *Let X be a finite CW complex, let $F \subset X$ be a compact subset and let $A_1 = QM_m(C(X))Q$, where $m \ge 1$ is an integer and $Q \in M_m(C(X))$ is a projection. Let $A = PM_m(C(F))P$, where $P = s(Q)$ and $s : A_1 \to A$ is the surjective homomorphism induced by the quotient map $C(X) \to C(F)$.*

Then, for any $\epsilon > 0$, any finite subset $\mathcal{F} \subset A$ and a non-decreasing map $\Delta : (0,1) \to (0,1)$, there exists $\delta > 0$, a finite subset $\mathcal{G} \subset A$ and a finite subset $\mathcal{P} \subset \underline{K}(A)$ satisfying the following:

Suppose that B is a unital separable simple C^*-algebra with tracial rank zero, $h : A \to B$ is a unital monomorphism and $u \in B$ is a unitary such that $\mu_{\tau\circ h}$ is Δ-distributed for all $\tau \in T(B)$,

$$\|[h(a), u]\| < \delta \textit{ for all } a \in \mathcal{G} \textit{ and } \mathrm{Bott}(h,u)|_{\mathcal{P}} = 0. \qquad \text{(e 17.27)}$$

Then, there exists a continuous rectifiable path of unitaries $\{u_t : t \in [0,1]\}$ of B such that

$$u_0 = u,\ u_1 = 1_B \textit{ and } \|[h(a), u_t]\| < \epsilon \textit{ for all } a \in \mathcal{F} \textit{ and } t \in [0,1]. \qquad \text{(e 17.28)}$$

Moreover,

$$\mathrm{Length}(\{u_t\}) \le 2\pi + \epsilon. \qquad \text{(e 17.29)}$$

Proof. It is clear that we may reduce the general case to the case that X is connected. So we now assume that X is connected. Let d be the dimension of X. By 8.15 of [**16**], there is a projection $e' \in M_{d+1}(QM_m(C(X))Q)$ such that $e' + Q$ is a trivial projection. So

$$(e' + Q)M_{d+1}(C(X))(e' + Q) \cong M_r(C(X))$$

for some integer $r \ge 1$. Put $e = s(e')$. It follows that

$$(e + P)M_{d+1}(C(F))(e + P) \cong M_r(C(F)).$$

Note that $P = 1_A$. Put $C = (e + 1_A)M_{d+1}(C(F))(e + 1_A)$. Note also that

$$\underline{K}(PCP) = \underline{K}(A).$$

If $h : A \to B$ is a unital homomorphism, then it extends to a unital homomorphism $h_1 : M_{d+1}(A) \to M_{d+1}(B)$. Let $h' = (h_1)|_C$. For any $\delta_1 > 0$ and a finite subset $\mathcal{G}_1 \subset A$, there exists $\delta_2 > 0$ and a finite subset $\mathcal{G}_2$ satisfying the following: if $u' \in B$ is a unitary such that

$$\|[h(f), u']\| < \delta_2 \text{ for all } f \in \mathcal{G}_2,$$

then

$$\|[h_1(g), U']\| < \delta_1 \text{ for all } \mathcal{G}_1'$$

where $U' = \mathrm{diag}(u', u', ..., u')$ and where $\mathcal{G}_1' = \{(g_{i,j}) : g \in \mathcal{G}_1\}$. In particular, with larger $\mathcal{G}_1$,

$$\|[1_C, U']\| < \delta_1.$$

Thus there is a unitary $V \in h_1(1_C)Bh_1(1_C)$ such that

$$\|V - h_1(1_C)U'h_1(1_C))\| < 2\delta_1 \text{ and } \|[h'(g), V]\| < 4\delta_1 \text{ for all } g \in \mathcal{G}_1' \cap C.$$

Let $\{e_{i,j}\}$ be a system of matrix units for M_r and be elements in $C \cong M_r(C(F))$.

Therefore, by applying 17.2 and its proof, for any $\epsilon_1 > 0$ and any $\mathcal{F}_1 \subset C(F)$, there is $\delta > 0$ and a finite subset $\mathcal{G} \subset A$ and a finite subset $\mathcal{P} \subset \underline{K}(A)$ satisfying the following: suppose that h and u satisfy the assumption of the lemma for the above δ and $\mathcal{G}$ and $\mathcal{P}$, there exists a path of unitaries $\{w_t : t \in [0,1]\}$ in $h'(e_{1,1})Bh'(e_{1,1})$ such that

$$\|w_0 - w\| < \epsilon_1/(4r)^2, \quad w_1 = e_{1,1}, \quad \|[h'(g), W_t]\| < \epsilon_1 \tag{e 17.30}$$

for all $t \in [0,1]$ and $g \in \mathcal{F}_1'$, and

$$\mathrm{Length}(\{w_t\}) \le 2\pi + \epsilon_1/2, \tag{e 17.31}$$

where w is a unitary in $e_{1,1}Ce_{1,1}$ with

$$\|w - e_{1,1}U'e_{1,1}\| < \epsilon_1/(2r)^2, \tag{e 17.32}$$

$W_t = \mathrm{diag}(w_t, w_t, ..., w_t) : t \in [0,1]$ and

$$\mathcal{F}_1' = \{(g_{i,j}) \in M_r(C(F)) \cong C : g_{i,j} \in \mathcal{F}_1\}.$$

We may assume that $P = 1_A \in \mathcal{F}_1$. For any finite subset $\mathcal{F} \subset A$, we may assume that $\mathcal{F} \subset \mathcal{F}_1'$, if $\mathcal{F}_1$ is sufficiently large. Note that $h'(g)P = Ph'(g)$ for all $g \in C$ (since h commutes with P). Note also that

$$\mathrm{Length}(\{W_t\}) \le 2\pi + \epsilon_1/4. \tag{e 17.33}$$

With sufficiently small ϵ_1 and sufficiently large $\mathcal{F}_1$, by applying 17.4, we obtain a path of unitaries $\{v_t : t \in [0,1]\}$ of B such that

$$\|v_0 - u\| < \epsilon/8, \quad v_1 = 1_B, \quad \|[h(f), v_t]\| < \epsilon/2 \text{ and} \tag{e 17.34}$$

$$\mathrm{Length}(\{v_t\}) \le 2\pi + \epsilon/2. \tag{e 17.35}$$

The lemma then follows by connecting v_0 to u with length no more than $\epsilon/2$. □

Lemma 17.6. *Let X be a finite CW complex, let $F \subset X$ be a compact subset and let $A_1 = QM_m(C(X))Q$, where $m \ge 1$ is an integer and $Q \in M_m(C(X))$ is a projection. Let $A = PM_m(C(F))P$, where $P = s(Q)$ and $s : A_1 \to A$ is the surjective homomorphism induced by the quotient map $C(X) \to C(F)$.*

Then, for any $\epsilon > 0$, any finite subset $\mathcal{F} \subset A$, there exists $\delta > 0$, a finite subset $\mathcal{G} \subset A$ and a finite subset $\mathcal{P} \subset \underline{K}(A)$ satisfying the following:

Suppose that B is a unital purely infinite simple C^*-algebra, $h : A \to B$ is a unital monomorphism and $u \in B$ is a unitary such that

$$\|[h(a), u]\| < \delta \text{ for all } a \in \mathcal{G} \text{ and } \mathrm{Bott}(h, u)|_{\mathcal{P}} = 0. \tag{e 17.36}$$

Then, there exists a continuous rectifiable path of unitaries $\{u_t : t \in [0,1]\}$ of B such that

$$u_0 = u, \quad u_1 = 1_B \text{ and } \|[h(a), u_t]\| < \epsilon \text{ for all } a \in \mathcal{F} \text{ and } t \in [0,1]. \tag{e 17.37}$$

Moreover,

$$\mathrm{Length}(\{u_t\}) \le 2\pi + \epsilon\pi. \tag{e 17.38}$$

PROOF. The proof is virtually identical to that of 17.5 but apply 17.3 instead of 17.2. □

Definition 17.7. Let A be a unital AH-algebra. Then A has the form $A = \lim_{n\to\infty}(A_n, \phi_n)$, where $A_n = \bigoplus_{i=1}^{r(n)} P_{i,n}M_{k(i,n)}(C(X_{i,n}))P_{i,n}$ and $P_{i,n} \in M_{k(i,n)}(C(X_{i,n}))$ is a projection and $X_{i,n}$ is a path connected finite CW complex. For convenience, without loss of generality, we may assume that each ϕ_n is unital. By replacing $X_{i,n}$ by a compact subset of $X_{i,n}$, we may assume that ϕ_n is injective for each n (but each $X_{i,n}$ is a compact subset of a path connected finite CW complex).

Let B be a unital C^*-algebra with a tracial state $\tau \in T(B)$ and $\psi : A \to B$ be a positive linear map. Define τ_n by $\tau_n = \tau\circ\psi\circ\phi_n$. Define μ_{τ_n} as in (iii) of 17.1. Let $\Delta_n : (0,1) \to (0,1)$ be an increasing map. We say that $\tau\circ\psi$ is $\{\Delta_n\}$-distributed if μ_{τ_n} is Δ_n -distributed (see 7.1).

Now if we further assume that A is simple, since now ϕ_n is assumed to be injective (where each $X_{k,n}$ is a compact subset of a connected finite CW complex), similar to 7.2, there exists an increasing map $\Delta_n : (0,1) \to (0,1)$ such that $\mu_{\tau\circ\phi_n}$ is Δ_n-distributed for all $\tau \in T(A)$. Moreover, if $\psi : A \to B$ is a unital homomorphism (necessarily injective since A is simple), then $t\circ\psi$ is $\{\Delta_n\}$-distributed for all $t \in T(B)$.

Theorem 17.8. *Let A be a unital AH-algebra, let $\epsilon > 0$ and $\mathcal{F} \subset A$ be a finite subset. Let $\Delta_n : (0,1) \to (0,1)$ be a sequence of increasing maps.*

Then there exists $\delta > 0$, a finite subset $\mathcal{G} \subset A$ and a finite subset $\mathcal{P} \subset \underline{K}(A)$ satisfying the following: Suppose that B is a unital separable simple C^*-algebra with tracial rank zero, $h : A \to B$ is a unital monomorphism, and suppose that there is a unitary $u \in B$ such that

$$\|[h(a), u]\| < \delta \text{ for all } f \in \mathcal{G}, \mathrm{Bott}(h,u)|_{\mathcal{P}} = 0 \text{ and} \tag{e 17.39}$$

$\mu_{\tau\circ h}$ is $\{\Delta_n\}$-distributed for all $\tau \in T(B)$. Then there exists a continuous path of unitaries $\{u_t : t \in [0,1]\}$ such that

$$u_0 = u,\ \ u_1 = u,\ \ \|[h(a), v_t]\| < \epsilon \text{ for all } f \in \mathcal{F} \text{ and } t \in [0,1]. \tag{e 17.40}$$

Moreover,

$$\|u_t - u_{t'}\| \le (2\pi+\epsilon)|t-t'| \text{ for all } t, t' \in [0,1] \tag{e 17.41}$$

$$\text{and } \mathrm{Length}(\{u_t\}) \le 2\pi + \epsilon. \tag{e 17.42}$$

PROOF. Let $\epsilon > 0$ and $\mathcal{F} \subset A$ be given. Write $A = \lim_{n\to\infty}(A_n, \phi_n)$, where each $A_n = \bigoplus_{i=1}^{r(n)} P_{i,n}M_{k(i)}(C(X_{i,n}))P_{i,n}$, where each $X_{i,n}$ is a connected finite CW complex. By replacing $X_{i,n}$ by one of its compact subset $F_{i,n}$, we may write $A = \overline{\cup_n A_n}$. Thus, without loss of generality, we may assume that $\mathcal{F} \subset A_n$ for some integer $n \ge 1$. We then apply 17.5.

Note to get (e 17.41), we can apply 9.3. □

Corollary 17.9. *Let A be a unital AH-algebra, let $\epsilon > 0$ and $\mathcal{F} \subset A$ be a finite subset. Suppose that B is a unital separable simple C^*-algebra with tracial rank zero and $h : A \to B$ is a unital monomorphism. Then there exists $\delta > 0$, a finite subset $\mathcal{G} \subset A$ and a finite subset $\mathcal{P} \subset \underline{K}(A)$ satisfying the following: Suppose that there is a unitary $u \in B$ such that*

$$\|[h(a), u]\| < \delta \text{ for all } f \in \mathcal{G} \text{ and } \mathrm{Bott}(h,u)|_{\mathcal{P}} = 0. \tag{e 17.43}$$

Then there exists a continuous path of unitaries $\{u_t : t \in [0,1]\}$ *such that*

$$u_0 = u, \ \ u_1 = u, \|[h(a), v_t]\| < \epsilon \text{ for all } f \in \mathcal{F} \textit{ and } t \in [0,1],$$
$$\|u_t - u_{t'}\| \le (2\pi + \epsilon)|t - t'| \textit{ for all } t, t' \in [0,1] \textit{ and}$$
$$\text{Length}(\{u_t\}) \le 2\pi + \epsilon.$$

Theorem 17.10. *Let A be a unital AH-algebra, let $\epsilon > 0$ and $\mathcal{F} \subset A$ be a finite subset. Then there exists $\delta > 0$, a finite subset $\mathcal{G} \subset A$ and a finite subset $\mathcal{P} \subset \underline{K}(A)$ satisfying the following: Suppose that B is a unital purely infinite simple C^*-algebra and $h : A \to B$ is a unital monomorphism and suppose that there is a unitary $u \in B$ such that*

$$\|[h(a), u]\| < \delta \textit{ for all } f \in \mathcal{G} \textit{ and } \text{Bott}(h, u)|_{\mathcal{P}} = 0. \tag{e 17.44}$$

Then there exists a continuous path of unitaries $\{u_t : t \in [0,1]\}$ *such that*

$$u_0 = u, \ \ u_1 = u, \|[h(a), v_t]\| < \epsilon \textit{ for all } f \in \mathcal{F} \textit{ and for all } t \in [0,1]$$

and

$$\|u_t - u_{t'}\| \le (2\pi + \epsilon)|t - t'| \textit{ for all } t, t' \in [0,1].$$

Consequently,

$$\text{Length}(\{u_t\}) \le 2\pi + \epsilon.$$

Proof. The proof follows the same argument used in 17.8 but applying 17.6. □

The additional significance of the following theorem is that, when A is also assumed to be simple, the "measure-distribution" is not needed in the assumption.

Theorem 17.11. *Let A be a unital simple AH-algebra. For any $\epsilon > 0$ and any finite subset $\mathcal{F} \subset A$, there exists $\delta > 0$, a finite subset $\mathcal{G} \subset A$ and a finite subset $\mathcal{P} \subset \underline{K}(A)$ satisfying the following: Suppose that B is a unital simple C^*-algebra of tracial rank zero, or A is a purely infinite simple C^*-algebra, suppose that $h : A \to B$ is a unital homomorphism and suppose that there is a unitary $u \in B$ such that*

$$\|[h(a), u]\| < \delta \textit{ and } \text{Bott}(h, u)|_{\mathcal{P}} = 0.$$

Then there exists a rectifiable continuous path of unitaries $\{u_t : t \in [0,1]\}$ *such that*

$$u_0 = u, \ \ u_1 = 1_B, \|[h(a), u_t]\| < \epsilon \text{ for all } f \in \mathcal{F}$$

and

$$\|u_t - u_{t'}\| \le (2\pi + \epsilon)|t - t'| \textit{ for all } t, t' \in [0,1].$$

Consequently

$$\text{Length}(\{u_t\}) \le 2\pi + \epsilon.$$

Proof. We will prove the case that B is a unital separable simple C^*-algebra with tracial rank zero. The case that B is a unital purely infinite simple C^*-algebra follows from 17.10.

From the discussion after 17.7, since A is a unital simple C^*-algebra, there is a sequence $\Delta_n : (0,1) \to (0,1)$ such that $\tau \circ \psi$ is $\{\Delta_n\}$-distributed for all $\tau \in T(A)$.

For any unital C^*-algebra B and any $t \in T(B)$ and any unital homomorphism $h : A \to B$, $t \circ h = \tau$ for some $\tau \in T(A)$. Thus $t \circ h$ is $\{\Delta_n\}$-distributed. In other words, once A is known, $\{\Delta_n\}$ can be determined independently of B and h. Thus the Theorem follows from 17.8. □

18. Concluding remarks

18.1. Suppose that $X = \sqcup_{i=1}^n X_i$, where each X_i is a connected compact metric space. We can define

$$\begin{aligned} \underline{L}_p(X) &= \max_{1\le i\le n} \underline{L}_p(X_i), \\ \overline{L}_p(X) &= \max_{1\le i\le n} \overline{L}_p(X_i) \text{ and} \\ L(X) &= \max_{1\le i\le n} L(X_i). \end{aligned}$$

A modification of the proofs of 13.5 shows the following version of Theorem 13.5:

Theorem 18.2. *Let X be a compact metric space with finitely many path connected components and let A be a unital purely infinite simple C^*-algebra. Suppose that $h_1, h_2 : C(X) \to A$ are two unital homomorphisms such that*

$$[h_1] = [h_2] \text{ in } KL(C(X), A).$$

Then, for any $\epsilon > 0$ and any finite subset $\mathcal{F} \subset C(X)$, there exist two unital homomorphisms $H_1, H_2 : C(X) \to C([0,1], A)$ such that

$$\pi_0 \circ H_1 \approx_{\epsilon/3} h_1, \pi_1 \circ H_1 \approx_{\epsilon/3} \pi_0 \circ H_2 \text{ and } \pi_1 \circ H_2 \approx_{\epsilon/3} h_2 \text{ on } \mathcal{F}.$$

Moreover,

$$\begin{aligned} \overline{\text{Length}}(\{\pi_t \circ H_1\}) &\le L(X)(1+2\pi) + \epsilon/2 \text{ and} \\ \overline{\text{Length}}(\{\pi_t \circ H_2\}) &\le L(X) + \epsilon/2. \end{aligned} \tag{e 18.1}$$

Similar statements also hold for Theorem 13.8, Theorem 13.9, Theorem 14.1, Theorem 14.2, Theorem 14.3, Theorem 15.1, Theorem 15.2 and Theorem 16.2.

For example, 13.8 holds for a general finite CW complex X :

Theorem 18.3. *Let X be a finite CW complex and let A be a unital separable simple C^*-algebra with tracial rank zero. Suppose that $\psi_1, \psi_2 : C(X) \to A$ are two unital monomorphisms such that*

$$[\psi_1] = [\psi_2] \text{ in } KL(C(X), A). \tag{e 18.2}$$

Then, for any $\epsilon > 0$ and any finite subset $\mathcal{F} \subset C(X)$, there exists a unital homomorphism $H : C(X) \to C([0,1], A)$ such that

$$\pi_0 \circ H \approx_\epsilon \psi_1 \text{ and } \pi_1 \circ H \approx_\epsilon \psi_2 \text{ on } \mathcal{F}. \tag{e 18.3}$$

Moreover, each $\pi_t \circ H$ is a monomorphism and

$$\overline{\text{Length}}(\{\pi_t \circ H\}) \le L(X) + \underline{L}_p(X) 2\pi + \epsilon. \tag{e 18.4}$$

18.4. One may also define $L(X)$ for general compact metric spaces. There are at least two possibilities. One may consider the following constant:

$$\sup\{ \text{Length}(\{\gamma\}) : \text{all rectifiable continuous paths in } X\}.$$

If $X = \lim_{\leftarrow n} X_n$, where each X_n is a finite CW complex, then one may consider the constant:

$$\inf \sup\{L(X_n) : n \in \mathbb{N}\},$$

where the infimum is taken among all possible reverse limits.

It seems that the second constant is easy to handle for our purpose. However, in general, one should not expect such a constant to be finite.

18.5. The last section presents some versions of the Basic Homotopy Lemma for AH-algebras. It seems possible that more general situation could be discussed. More precisely, for example, one may ask if 17.9 holds for a general separable amenable C^*-algebra A. However, it is beyond the scope of this research to answer this question. While we know several cases are valid, in general, further study is required.

There are several valid versions of the Super Homotopy Lemma for non-commutative cases and could be presented here. However, given the length of this work, we choose not include them.

18.6. Applications of the results in this work have been mentioned in the introduction. It is certainly desirable to present some of them here. Nevertheless, the length of this work again prevents us from including them here without further prolonging the current work. We would like however, to mention that, for example, Theorem 17.8 and Theorem 17.9 are used in [**34**] and [**37**] where we show crossed products of unital AH-algebras by $\mathbb{Z}$, and by finitely generated abelian groups can be embedded into unital simple AF-algebras if and only if the crossed products have faithful tracial states, respectively. It also plays an essential role in the recent development in the classification of amenable simple C^*-algebras (see [**35**], [**55**], [**36**] and [**39**]).

Bibliography

[1] C. Akemann and F. W. Shultz, *Perfect C^*-algebras*, Mem. Amer. Math. Soc. **55** (1985), no. 326, xiii+117 pp. MR787540 (87h:46117)

[2] B. Blackadar and M. Rørdam, *Extending states on preordered semigroups and the existence of quasitraces on C^*-algebras*, J. Algebra **152** (1992), 240–247. MR1190414 (93k:46049)

[3] O. Bratteli, G. A. Elliott, D. E. Evans and A. Kishimoto, *Homotopy of a pair of approximately commuting unitaries in a simple C^*-algebra*, J. Funct. Anal. **160** (1998), 466–523. MR1665295 (99m:46132)

[4] N. Brown, *Invariant means and finite representation theory of C^*-algebras*, Mem. Amer. Math. Soc., **184** (2006), viii+105 pp. MR2263412 (2008d:46070)

[5] J. Cuntz, *K-theory for certain C^*-algebras*, Ann. of Math. **113** (1981), 181–197. MR604046 (84c:46058)

[6] M. Dădărlat, *Approximately unitarily equivalent morphisms and inductive limit C^*-algebras*, K-Theory **9** (1995), 117–137. MR1340842 (96h:46088)

[7] M. Dădărlat, *Morphisms of simple tracially AF algebras*, Internat. J. Math. **15** (2004), 919–957. MR2106154 (2005i:46061)

[8] M. Dădărlat and T. A. Loring, *The K-theory of abelian subalgebras of AF algebras*, J. Reine Angew. Math. **432** (1992), 39–55. MR1184757 (94b:46099)

[9] M. Dădărlat and T. Loring, *A universal multicoefficient theorem for the Kasparov groups*, Duke Math. J. **84** (1996), 355–377. MR1404333 (97f:46109)

[10] G. A. Elliott and M. Rørdam, *Classification of certain infinite simple C^*-algebras. II*, Comment. Math. Helv. **70** (1995), 615–638. MR1360606 (96e:46080b)

[11] R. Exel and T. Loring, *Invariants of almost commuting unitaries*, J. Funct. Anal. **95** (1991), 364–376. MR1092131 (92a:46083)

[12] R. Exel and T. Loring, *Almost commuting unitary matrices*, Proc. Amer. Math. Soc. **106** (1989), 913–915. MR975641 (89m:15003)

[13] G. Gong and H. Lin, *Almost multiplicative morphisms and almost commuting matrices*, J. Operator Theory **40** (1998), 217–275. MR1660385 (2000c:46105)

[14] G. Gong and H. Lin, *Almost multiplicative morphisms and K-theory*, Internat. J. Math. **11** (2000), 983–1000. MR1797674 (2001j:46081)

[15] S. Hu, H. Lin, Y. Xue, *Limits of homomorphisms with finite-dimensional range*, Internat. J. Math. **16** (2005), 807–821. MR2158961 (2006d:46067)

[16] D. Husemoller, *Fibre Bundles*, McGraw-Hill, New York, 1966, reprinted in Springer-Verlag Graduate Texts in Mathmatics. MR0229247 (37:4821)

[17] E. Kirchberg and N. C. Phillips, *Embedding of exact C^*-algebras in the Cuntz algebra $\mathcal{O}_2$*, J. Reine Angew. Math. **525** (2000), 17–53. MR1780426 (2001d:46086a)

[18] A. Kishimoto, *Automorphisms of* **AT** *algebras with the Rohlin property*, J. Operator Theory **40** (1998), 277–294. MR1660386 (99j:46073)

[19] A. Kishimoto and A. Kumjian, *The Ext class of an approximately inner automorphism. II*, J. Operator Theory **46** (2001), 99–122. MR1862181 (2003d:46081)

[20] L. Li, *C^*-algebra homomorphisms and KK-theory*, K-Theory **18** (1999), 161–172. MR1711712 (2000j:46130)

[21] H. Lin, *Exponential rank of C^*-algebras with real rank zero and the Brown-Pedersen conjectures*, J. Funct. Anal. **114** (1993), 1–11. MR1220980 (95a:46079)

[22] H. Lin, *Almost commuting selfadjoint matrices and applications*, Operator algebras and their applications (Waterloo, ON, 1994/1995), 193–233, Fields Inst. Commun., 13, Amer. Math. Soc., Providence, RI, 1997. MR1424963 (98c:46121)

[23] H. Lin, *On the classification of C^*-algebras of real rank zero with zero K_1*, J. Operator Theory **35** (1996), 147–178. MR1389648 (98g:46086)

[24] H. Lin, *Almost multiplicative morphisms and some applications*, J. Operator Theory **37** (1997),121–154. MR1438204 (98b:46091)

[25] H. Lin, *When almost multiplicative morphisms are close to homomorphisms*, Trans. Amer. Math. Soc. **351** (1999), 5027–5049. MR1603918 (2000j:46107)

[26] H. Lin, *Embedding an AH-algebra into a simple C^*-algebra with prescribed KK-data*, K-Theory **24** (2001), 135–156. MR1869626 (2002i:46053)

[27] H. Lin, *Tracial topological ranks of C^*-algebras*, Proc. London Math. Soc., **83** (2001), 199-234. MR1829565 (2002e:46063)

[28] H. Lin, *Classification of simple C^*-algebras and higher dimensional noncommutative tori*, Ann. of Math. (2) **157** (2003), 521–544. MR1973053 (2004b:46089)

[29] H. Lin, *An Introduction to the Classification of Amenable C^*-Algebras*, World Scientific Publishing Co., Inc., River Edge, NJ, 2001. xii+320 pp. ISBN: 981-02-4680-3. MR1884366 (2002k:46141)

[30] H. Lin, *Classification of simple C^*-algebras with tracial topological rank zero*, Duke Math. J. **125** (2004), 91-119. MR2097358 (2005i:46064)

[31] H. Lin, *Traces and simple C^*-algebras with tracial topological rank zero*, J. Reine Angew. Math. **568** (2004), 99–137. MR2034925 (2005b:46122)

[32] H. Lin, *A separable Brown-Douglas-Fillmore theorem and weak stability*, Trans. Amer. Math. Soc. **356** (2004), 2889–2925. MR2052601 (2005d:46116)

[33] H. Lin, *Classification of homomorphisms and dynamical systems*, Trans. Amer. Math. Soc. **359** (2007), 859-895. MR2255199 (2008d:46082)

[34] H. Lin, *AF-emebddings of crossed products of AH-algebras by $\mathbb{Z}$ and asymptotic AF-embeddings*, Indiana Univ. Math. J. **57** (2008), 891–944. MR2414337

[35] H. Lin, *Asymptotic unitary equivalence and asymptotically inner automorphisms*, Amer. J. Math., to appear, arXiv:math/0703610.

[36] H. Lin, *Localizing the Elliott Conjecture at Strongly Self-absorbing C^*-algebras –An Appendix*, preprint, arXiv:0709.1654.

[37] H. Lin, *AF-embedding of the crossed products of AH-algebras by finitely generated abelian groups*, Int. Math. Res. Pap. IMRP 2008, no. 3, Art. ID rpn007, 67 pp. MR2457848

[38] H. Lin, *Asymptotically unitary equivalence and classification of simple amenable C^*-algebras*, preprint, arXiv:0806.0636.

[39] H. Lin and Z. Niu, *Lifting KK-elements, asymptotical unitary equivalence and classification of simple C^*-algebras*, Adv. Math. **219** (2008), 1729–1769. MR2458153 (2009g:46118)

[40] H. Lin and N. C. Phillips, *Almost multiplicative morphisms and the Cuntz algebra $\mathcal{O}_2$*, Internat. J. Math. **6** (1995), 625–643. MR1339649 (97h:46097)

[41] H. Lin and H. Su, *Classification of direct limits of generalized Toeplitz algebras*, Pacific J. Math. **181** (1997), 89–140. MR1491037 (99h:46106)

[42] T. Loring, *Stable relations. II. Corona semiprojectivity and dimension-drop C^*-algebras*, Pacific J. Math. **172** (1996), 461–475. MR1386627 (97c:46070)

[43] T. Loring, *K-theory and asymptotically commuting matrices*, Canad. J. Math. **40** (1988), 197-216. MR928219 (89b:47022)

[44] S. Mardesic, *On covering dimension and inverse limits of compact spaces*, Illinois J. Math **4** (1960), 278–291. MR0116306 (22:7101)

[45] H. Matui, *AF embeddability of crossed products of AT algebras by the integers and its application*, J. Funct. Anal. **192** (2002), 562–580. MR1923414 (2003h:46101)

[46] N. C. Phillips, *Approximation by unitaries with finite spectrum in purely infinite C^*-algebras.*, J. Funct. Anal. **120** (1994), 98–106. MR1262248 (95c:46092)

[47] N. C. Phillips, *A classification theorem for nuclear purely infinite simple C^*-algebras*, Doc. Math. **5** (2000), 49–114. MR1745197 (2001d:46086b)

[48] M. Rørdam, *Classification of inductive limits of Cuntz algebras*, J. Reine Angew. Math. **440** (1993), 175–200. MR1225963 (94k:46120)

[49] M. Rørdam, *Classification of certain infinite simple C^*-algebras*, J. Funct. Anal. **131** (1995), 415–458. MR1345038 (96e:46080a)

[50] M. Rørdam, *Classification of Nuclear C^*-algebras*, Encyclopaedia of Mathematical Sciences **126**, Springer 2002. MR1878882 (2003i:46060)

[51] J. Rosenberg and C. Schochet, *The Künneth theorem and the universal coefficient theorem for Kasparov's generalized K-functor*, Duke Math. J. **55** (1987), 431–474. MR894590 (88i:46091)

[52] A. Toms and W. Winter, *Minimal Dynamics and K-theoretic Rigidity: Elliott's Conjecture*, preprint, arXiv:0903.4133.

[53] W. Winter, *On topologically finite-dimensional simple C^*-algebras*, Math. Ann. **332** (2005), 843–878. MR2179780 (2006i:46102)

[54] W. Winter, *On the classification of simple $\mathcal{Z}$-stable C^*-algebras with real rank zero and finite decomposition rank*, J. London Math. Soc. **74** (2006), 167–183. MR2254559 (2007g:46086)

[55] W. Winter, *Localizing the Elliott conjecture at strongly self-absorbing C^*-algebras*, preprint, arXiv:0708.0283.

[56] S. Zhang, *A Riesz decomposition property and ideal structure of multiplier algebras*, J. Operator Theory **24** (1990), 209–225. MR1150618 (93b:46116)

[57] S. Zhang, *Matricial structure and homotopy type of simple C^*-algebras with real rank zero*, J. Operator Theory **26** (1991), 283–312. MR1225518 (94f:46075)

Editorial Information

To be published in the *Memoirs*, a paper must be correct, new, nontrivial, and significant. Further, it must be well written and of interest to a substantial number of mathematicians. Piecemeal results, such as an inconclusive step toward an unproved major theorem or a minor variation on a known result, are in general not acceptable for publication.

Papers appearing in *Memoirs* are generally at least 80 and not more than 200 published pages in length. Papers less than 80 or more than 200 published pages require the approval of the Managing Editor of the Transactions/Memoirs Editorial Board. Published pages are the same size as those generated in the style files provided for $\mathcal{AMS}$-LaTeX or $\mathcal{AMS}$-TeX.

Information on the backlog for this journal can be found on the AMS website starting from `http://www.ams.org/memo`.

A Consent to Publish and Copyright Agreement is required before a paper will be published in the *Memoirs*. After a paper is accepted for publication, the Providence office will send a Consent to Publish and Copyright Agreement to all authors of the paper. By submitting a paper to the *Memoirs*, authors certify that the results have not been submitted to nor are they under consideration for publication by another journal, conference proceedings, or similar publication.

Information for Authors

Memoirs is an author-prepared publication. Once formatted for print and on-line publication, articles will be published as is with the addition of AMS-prepared frontmatter and backmatter. Articles are not copyedited; however, confirmation copy will be sent to the authors.

Initial submission. The AMS uses Centralized Manuscript Processing for initial submissions. Authors should submit a PDF file using the Initial Manuscript Submission form found at `www.ams.org/peer-review-submission`, or send one copy of the manuscript to the following address: Centralized Manuscript Processing, MEMOIRS OF THE AMS, 201 Charles Street, Providence, RI 02904-2294 USA. If a paper copy is being forwarded to the AMS, indicate that it is for *Memoirs* and include the name of the corresponding author, contact information such as email address or mailing address, and the name of an appropriate Editor to review the paper (see the list of Editors below).

The paper must contain a *descriptive title* and an *abstract* that summarizes the article in language suitable for workers in the general field (algebra, analysis, etc.). The *descriptive title* should be short, but informative; useless or vague phrases such as "some remarks about" or "concerning" should be avoided. The *abstract* should be at least one complete sentence, and at most 300 words. Included with the footnotes to the paper should be the 2010 *Mathematics Subject Classification* representing the primary and secondary subjects of the article. The classifications are accessible from `www.ams.org/msc/`. The Mathematics Subject Classification footnote may be followed by a list of *key words and phrases* describing the subject matter of the article and taken from it. Journal abbreviations used in bibliographies are listed in the latest *Mathematical Reviews* annual index. The series abbreviations are also accessible from `www.ams.org/msnhtml/serials.pdf`. To help in preparing and verifying references, the AMS offers MR Lookup, a Reference Tool for Linking, at `www.ams.org/mrlookup/`.

Electronically prepared manuscripts. The AMS encourages electronically prepared manuscripts, with a strong preference for $\mathcal{AMS}$-LaTeX. To this end, the Society has prepared $\mathcal{AMS}$-LaTeX author packages for each AMS publication. Author packages include instructions for preparing electronic manuscripts, samples, and a style file that generates the particular design specifications of that publication series. Though $\mathcal{AMS}$-LaTeX is the highly preferred format of TeX, author packages are also available in $\mathcal{AMS}$-TeX.

Authors may retrieve an author package for *Memoirs of the AMS* from `www.ams.org/journals/memo/memoauthorpac.html` or via FTP to `ftp.ams.org` (login as `anonymous`, enter your complete email address as password, and type `cd pub/author-info`). The

AMS Author Handbook and the *Instruction Manual* are available in PDF format from the author package link. The author package can also be obtained free of charge by sending email to tech-support@ams.org (Internet) or from the Publication Division, American Mathematical Society, 201 Charles St., Providence, RI 02904-2294, USA. When requesting an author package, please specify AMS-LaTeX or AMS-TeX and the publication in which your paper will appear. Please be sure to include your complete mailing address.

After acceptance. The source files for the final version of the electronic manuscript should be sent to the Providence office immediately after the paper has been accepted for publication. The author should also submit a PDF of the final version of the paper to the editor, who will forward a copy to the Providence office.

Accepted electronically prepared files can be submitted via the web at www.ams.org/submit-book-journal/, sent via FTP, or sent on CD-Rom or diskette to the Electronic Prepress Department, American Mathematical Society, 201 Charles Street, Providence, RI 02904-2294 USA. TeX source files and graphic files can be transferred over the Internet by FTP to the Internet node ftp.ams.org (130.44.1.100). When sending a manuscript electronically via CD-Rom or diskette, please be sure to include a message indicating that the paper is for the *Memoirs.*

Electronic graphics. Comprehensive instructions on preparing graphics are available at www.ams.org/authors/journals.html. A few of the major requirements are given here.

Submit files for graphics as EPS (Encapsulated PostScript) files. This includes graphics originated via a graphics application as well as scanned photographs or other computer-generated images. If this is not possible, TIFF files are acceptable as long as they can be opened in Adobe Photoshop or Illustrator.

Authors using graphics packages for the creation of electronic art should also avoid the use of any lines thinner than 0.5 points in width. Many graphics packages allow the user to specify a "hairline" for a very thin line. Hairlines often look acceptable when proofed on a typical laser printer. However, when produced on a high-resolution laser imagesetter, hairlines become nearly invisible and will be lost entirely in the final printing process.

Screens should be set to values between 15% and 85%. Screens which fall outside of this range are too light or too dark to print correctly. Variations of screens within a graphic should be no less than 10%.

Inquiries. Any inquiries concerning a paper that has been accepted for publication should be sent to memo-query@ams.org or directly to the Electronic Prepress Department, American Mathematical Society, 201 Charles St., Providence, RI 02904-2294 USA.

Editors

This journal is designed particularly for long research papers, normally at least 80 pages in length, and groups of cognate papers in pure and applied mathematics. Papers intended for publication in the *Memoirs* should be addressed to one of the following editors. The AMS uses Centralized Manuscript Processing for initial submissions to AMS journals. Authors should follow instructions listed on the Initial Submission page found at www.ams.org/memo/memosubmit.html.

Algebra, to ALEXANDER KLESHCHEV, Department of Mathematics, University of Oregon, Eugene, OR 97403-1222; e-mail: ams@noether.uoregon.edu

Algebraic geometry, to DAN ABRAMOVICH, Department of Mathematics, Brown University, Box 1917, Providence, RI 02912; e-mail: amsedit@math.brown.edu

Algebraic geometry and its applications, to MINA TEICHER, Emmy Noether Research Institute for Mathematics, Bar-Ilan University, Ramat-Gan 52900, Israel; e-mail: teicher@macs.biu.ac.il

Algebraic topology, to ALEJANDRO ADEM, Department of Mathematics, University of British Columbia, Room 121, 1984 Mathematics Road, Vancouver, British Columbia, Canada V6T 1Z2; e-mail: adem@math.ubc.ca

Combinatorics, to JOHN R. STEMBRIDGE, Department of Mathematics, University of Michigan, Ann Arbor, Michigan 48109-1109; e-mail: JRS@umich.edu

Commutative and homological algebra, to LUCHEZAR L. AVRAMOV, Department of Mathematics, University of Nebraska, Lincoln, NE 68588-0130; e-mail: avramov@math.unl.edu

Complex analysis and harmonic analysis, to ALEXANDER NAGEL, Department of Mathematics, University of Wisconsin, 480 Lincoln Drive, Madison, WI 53706-1313; e-mail: nagel@math.wisc.edu

Differential geometry and global analysis, to CHRIS WOODWARD, Department of Mathematics, Rutgers University, 110 Frelinghuysen Road, Piscataway, NJ 08854; e-mail: ctw@math.rutgers.edu

Dynamical systems and ergodic theory and complex analysis, to YUNPING JIANG, Department of Mathematics, CUNY Queens College and Graduate Center, 65-30 Kissena Blvd., Flushing, NY 11367; e-mail: Yunping.Jiang@qc.cuny.edu

Functional analysis and operator algebras, to DIMITRI SHLYAKHTENKO, Department of Mathematics, University of California, Los Angeles, CA 90095; e-mail: shlyakht@math.ucla.edu

Geometric analysis, to WILLIAM P. MINICOZZI II, Department of Mathematics, Johns Hopkins University, 3400 N. Charles St., Baltimore, MD 21218; e-mail: trans@math.jhu.edu

Geometric topology, to MARK FEIGHN, Math Department, Rutgers University, Newark, NJ 07102; e-mail: feighn@andromeda.rutgers.edu

Harmonic analysis, representation theory, and Lie theory, to E. P. VAN DEN BAN, Department of Mathematics, Utrecht University, P.O. Box 80 010, 3508 TA Utrecht, The Netherlands; e-mail: E.P.vandenBan@uu.nl

Logic, to STEFFEN LEMPP, Department of Mathematics, University of Wisconsin, 480 Lincoln Drive, Madison, Wisconsin 53706-1388; e-mail: lempp@math.wisc.edu

Number theory, to JONATHAN ROGAWSKI, Department of Mathematics, University of California, Los Angeles, CA 90095; e-mail: jonr@math.ucla.edu

Number theory, to SHANKAR SEN, Department of Mathematics, 505 Malott Hall, Cornell University, Ithaca, NY 14853; e-mail: ss70@cornell.edu

Partial differential equations, to GUSTAVO PONCE, Department of Mathematics, South Hall, Room 6607, University of California, Santa Barbara, CA 93106; e-mail: ponce@math.ucsb.edu

Partial differential equations and dynamical systems, to PETER POLACIK, School of Mathematics, University of Minnesota, Minneapolis, MN 55455; e-mail: polacik@math.umn.edu

Probability and statistics, to RICHARD BASS, Department of Mathematics, University of Connecticut, Storrs, CT 06269-3009; e-mail: bass@math.uconn.edu

Real analysis and partial differential equations, to DANIEL TATARU, Department of Mathematics, University of California, Berkeley, Berkeley, CA 94720; e-mail: tataru@math.berkeley.edu

All other communications to the editors, should be addressed to the Managing Editor, ROBERT GURALNICK, Department of Mathematics, University of Southern California, Los Angeles, CA 90089-1113; e-mail: guralnic@math.usc.edu.

Titles in This Series

For a complete list of titles in this series, visit the
AMS Bookstore at **www.ams.org/bookstore/**.